Skin and Facade　Materials　Structure　Shape　Detail
表皮与立面　　　材料　　　结构　　造型　　细部

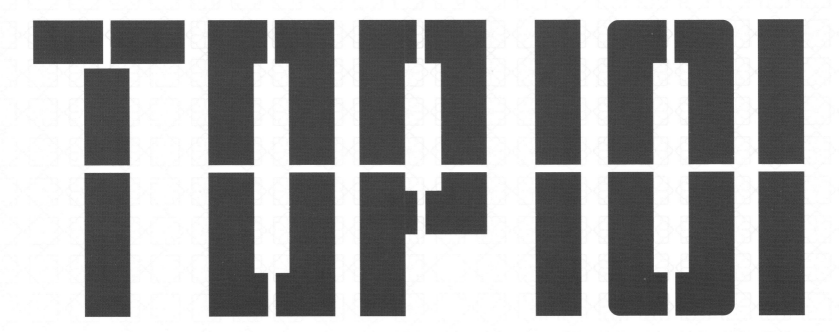

World's New Buildings III
101 世界最佳新建筑　佳图文化 编

中国林业出版社

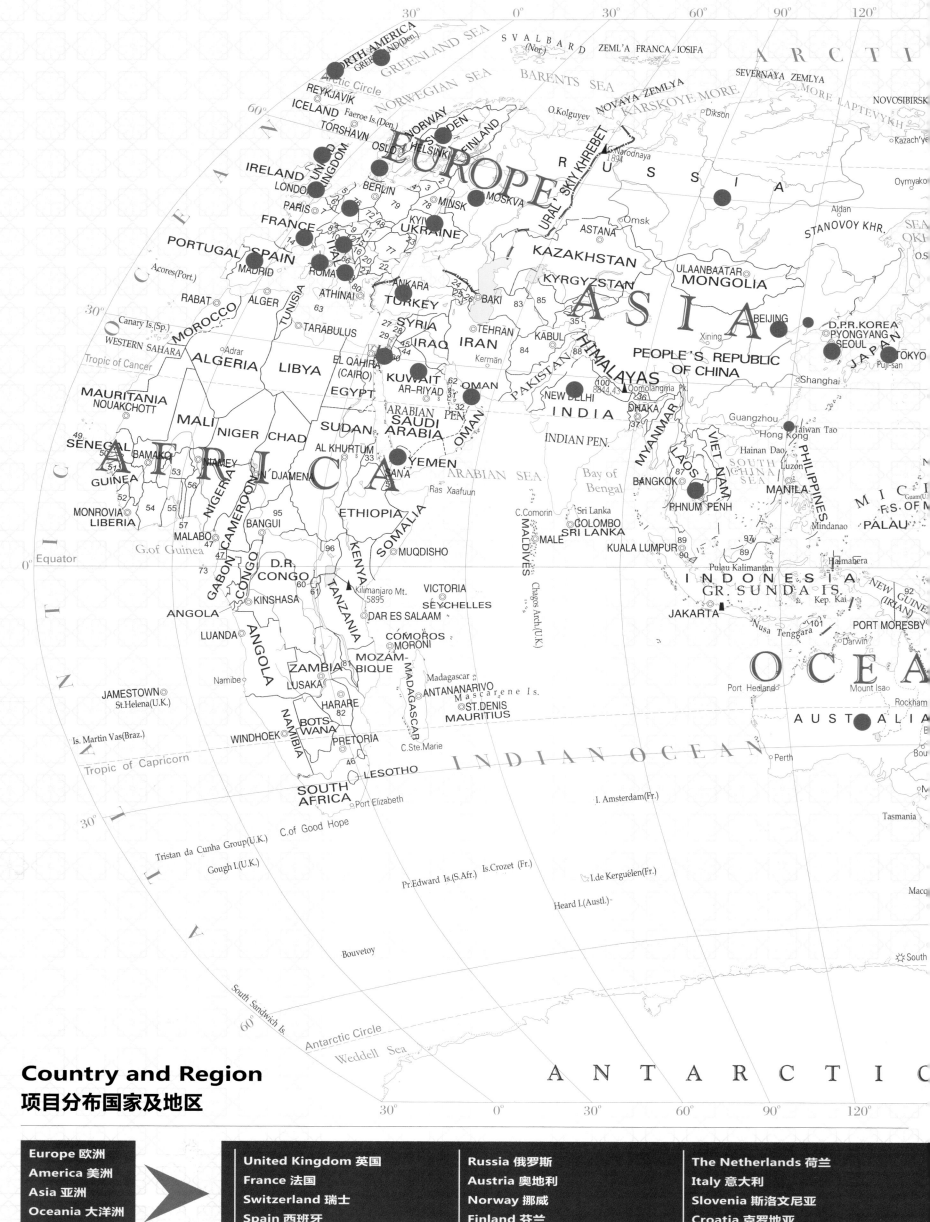

Country and Region
项目分布国家及地区

- Europe 欧洲
- America 美洲
- Asia 亚洲
- Oceania 大洋洲
- Africa 非洲

→

- United Kingdom 英国
- France 法国
- Switzerland 瑞士
- Spain 西班牙
- Russia 俄罗斯
- Austria 奥地利
- Norway 挪威
- Finland 芬兰
- The Netherlands 荷兰
- Italy 意大利
- Slovenia 斯洛文尼亚
- Croatia 克罗地亚

图书在版编目（CIP）数据

101世界最佳新建筑Ⅲ：汉英对照 / 佳图文化 主编． -- 北京 ：中国林业出版社，2013.7

ISBN 978-7-5038-7044-6

Ⅰ．①1… Ⅱ．①佳… Ⅲ．①建筑设计－图集②景观设计－图集 Ⅳ．① TU206 ② TU986.2-64

中国版本图书馆CIP数据核字（2013）第 089583 号

主编：佳图文化
策划：王志

中国林业出版社·建筑与家居图书出版中心
责任编辑：李 顺　唐 杨
出版咨询：（010）83223051

出　版：中国林业出版社（100009 北京西城区德内大街刘海胡同7号）
网　站：http://lycb.forestry.gov.cn/
印　刷：利丰雅高印刷（深圳）有限公司
发　行：中国林业出版社
电　话：（010）83224477
版　次：2013年7月第1版
印　次：2013年7月第1次
开　本：889mm×1194mm 1 / 8
印　张：60
字　数：400 千字
定　价：798 .00元

Preface 前言

As a new sequel of the bestseller - TOP 101 World's New Buildings I and II, the book has been greatly recommended by many world-class architectural institutes. The book focuses on the new buildings of the whole world with high starting point and wide scope. It has selected the latest great projects from Italy, Russia, Spain, France, UK, Switzerland, The Netherland, Norway, Austria, Croatia and Slovenia of Europe; USA, Canada and Columbia of America; China, Japan, South Korea, Turkey, Saudi Arabia, the Philippines and Indonesia of Asia.

There are six chapters, including Shopping Mall, Office Building, Hotel Building, Transportation and Sports Building and Art Building, etc. All these public buildings are innovative and representative, attached with technical drawings such as site plans, layouts, plans, elevations, sections, CAD construction drawings, detail drawings, renderings, and 3D models. Also each project is exhibited with professional built photos, notes and diagram form. In addition, the book has described these projects from various aspects such as keywords and features, overview, façade and skin, shape and structure and materials, which will be a great reference for the workers in the architecture industry.

As a masterpiece reflecting the latest global buildings, we try to respect the originals with the ideas of innovation, ecology, and energy conservation to ensure it is scientific, useful and perspective.

　　本书是继《101世界最佳新建筑》前两本在国内外热销后的全新续篇，受到世界各地顶级建筑机构的热烈好评及大力支持。全书放眼全球，起点高，涉足领域广，精选了来自欧洲的意大利、俄罗斯、西班牙、法国、英国、瑞士、荷兰、挪威、芬兰、奥地利、克罗地亚、斯洛文尼亚，美洲的美国、加拿大、哥伦比亚及亚洲的中国、日本、韩国、土耳其、沙特阿拉伯、菲律宾、印尼等国家和地区的最新的代表案例。

　　全书共分6大章节，包括大型购物中心、办公建筑、酒店建筑、交通体育建筑、医疗建筑以及艺术建筑等。书中展现了极具创新精神又具有代表性的最新公共建筑佳作，大量的总平面图、规划图、平面图、立面图、剖面图、CAD施工图、节点详图、效果图、三维模型图等珍贵详细的技术图纸，同时每个案例都精选专业的建成实景图，另还配以注解和图示；此外，本书从项目关键词与亮点、项目概况、立面与表皮、造型与结构以及材料运用等各方面进行精炼概括和提炼，相信这些必将为建筑工作者提供参考借鉴，激发其创作灵感。

　　作为一本整体反映当今全球范围内新建筑的代表性精品大作，我们力求做到尊重原创，创新、生态、节能概念至上，以确保本书的科学性、实用性和前瞻性。

Facade Material Index
立面材料索引

1 Metal Facade 金属立面

Aalta / Aalta 办公楼	096
ME Barcelona Hotel / 西班牙巴塞罗那 ME 酒店	206
CTA Morgan Street Station / CTA 摩根车站	264
Sports Palace / 皇家体育场	314
Museum MUMAC Museum of Coffee Machine / MUMAC 咖啡机博物馆	432
Tampa Museum of Art Tampa, Florida / 佛罗里达坦帕美术馆	460

2 Tile Facade 砖材立面

Ahlström Salmisaari Office Building / Ahlström Salmisaari 办公楼	104
Mint Hotel Amsterdam / 阿姆斯特丹 Mint 酒店	252

3 Stone Façade 石材立面

A Canadian Museum in a ChurchClaire and Marc Bourgie Pavilion of Quebec and Canadian Art / 蒙特利尔艺术博物馆：古老教堂中的艺术博物馆	400
Taiyuan Museum of Art / 太原美术馆	452

4 Concrete Façade 混凝土立面

Steel Band / Steel Band 办公楼	088
O-14 / O-14 商业大厦	174
Tel Aviv Museum of Art / 特拉维夫艺术博物馆	422

5 Glass Facade 玻璃立面

Interspar Fürstenfeld / Interspar Fürstenfeld 超市	026
Metropolis / 大都会	040
Mecenatpolis / Mecenatpolis	056
Herma Parking Building / 赫尔玛停车大楼	070
Westraven / Westraven 办公楼	112
Head office of Quebecor / 魁北克总部	134

Centennial Place / 世纪广场	152
Exaltis Office Tower / Exaltis 办公大楼	160
Tempo Office Tower / 节奏办公大楼	168
Drexel Parking/Office Building / 德雷克塞尔办公楼及停车场	188
Zuellig Building / 裕利大厦	194
Fukoku Tower / 大阪富国大厦	198
NH-Fieramilano Hotel / 米兰国际展览中心 NH 酒店	218
The Sheraton Milan Malpensa Airport Hotel & Conference Centre / 米兰马尔彭萨机场喜来登酒店及会议中心	240
Mall of Asia Arena / 亚洲商城体育馆	322
Children's Hospital Colorado / 科罗拉多州儿童医院	360
Laboratories, Haukeland University Hospital / Haukeland 大学医院实验室大楼	366
National Cowboy and Western Heritage Museum / 美国西部牛仔历史博物馆	408
Lightcatcher at the Whatcom Museum / 沃特科姆博物馆莱特卡彻楼	416

6 Wooden Facade 木质立面

Alésia Archaeological Museum / 阿莱西亚考古博物馆	390
Kamppi Chapel / 康比教堂	442

7 Plastic Facade 塑料立面

Olympic Shooting Venue / 伦敦奥运射击场	284

8 Mixed Facade 混合立面

Pendorya Shopping Center / PENDORYA 购物中心	032
Dot Envelope / Dot Envelope 商场	048
Skiterminal – Schladming / 施拉德明滑雪度假综合体	064
Manitoba Hydro Place / 马尼托巴水电局大楼	144
Hotel Well / Well 酒店	226
Hotel in Mengibar / 门希瓦尔酒店	234
ÖBB Rail Service Center / OBB 铁路服务中心	270
FK Austria Wien Training Academy / 奥地利维也纳 FK 足球训练学校	276
Four Sport Scenarios / 四个体育场	292
Richmond Olympic Oval / 列治文奥运速度滑冰馆	328
Olympic Tennis Centre, Madrid, Spain / 西班牙马德里奥林匹克网球中心	334
Expansion of Centre Sportif J.C. Malépart / J.C. Malépart 体育中心扩建项目	348
Diagonal Clinic / 西班牙 Diagonal 诊所	374
Kochi Health Sciences Center / 高知健康科学中心	382

The facade materials Introduction and analysis 立面材料简介及分析

Metal Facade 金属立面

In metal wall or roof system, metal materials are used in stead of traditional building materials. With strict standards for the production of metal panels as well as for the assembly techniques of the wall and roof system, metal materials are widely promoted and applied. For the facade design of buildings, metal materials such as metal panel, perforated metal, wire mesh, metal grill, clad metal sheet, metal sandwich panel, woven wire mesh and so on.

金属墙面以及屋面系统是使用金属材料替代传统材料建造建筑物墙面以及屋面的一项建筑技术。无论是金属板材的生产还是墙体屋面的成套技术均有多项严格的标准，这对于金属材料的运用和推广起到了重要的作用。在建筑立面中，最常用的金属制品有金属板、穿孔金属板、金属网、金属格栅、复合金属薄板、金属夹芯板、金属编织板等。

Metal Facade 金属板

In metal wall or roof system, metal materials are used in stead of traditional building materials. With strict standards for the production of metal panels as well as for the assembly techniques of the wall and roof system, metal materials are widely promoted and applied. For the facade design of buildings, metal materials such as metal panel, perforated metal, wire mesh, metal grill, clad metal sheet, metal sandwich panel, woven wire mesh and so on.

该材料具有较好的耐磨性。金属覆面体系在近年来被广泛的应用，这种体系易于连接，通常用隐形的卡子固定在支撑的结构上。预制金属板可以创造出精致的立面效果，尤其从远处看，呈现出不同肌理的金属表皮外观。这是在建筑外立面中运用非常广泛的一种金属制品。

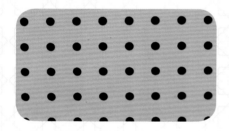

Perforated Metal 穿孔金属板

Perforated metal is made through the metal stamping and sheet metal manufacturing process. The perforation diameter can be less than 1mm or long to 500mm, and the distances between holes can be different. The holes can be perforated in straight line or out-of-line, in round, square or other shapes. Perforated materials are very versatile. For lightweight decorative elements to load-bearing structural components, perforated metal offers unique opportunities to combine strength, functionality and beauty. Perforated metal is also an excellent means of achieving sustainable design objectives.

这种金属板最主要的特点是在金属薄板上通过电脑控制打孔，穿孔直径可以小于1mm，大的孔径可达500mm，并且可以有不同的间距。穿孔可以沿着直线或者采用偏移的方式，可以是圆的、方的或者是其他不同的装饰性的形式。穿孔金属板易于加工，生产方式高度工业化，具有丰富的表面肌理，使其成为一种经济的、通用性的建筑材料。

Wire Mesh 金属网

Wire mesh is usually used for facing. The manufacturing process will not bring any waste. Metals such as iron, steel, aluminum and lightweight alloy as well as cooper, nickel and zinc can be processed by cutting or stretching for wire mesh. This product will be used as the translucent door, or used in curtain wall and facade.

该金属饰面的生产过程不造成任何的浪费，仅仅是对于薄板切割或者张拉成型，这个过程适用于铁、钢、铝和轻质合金，以及铜、镍、锌等。该材料具有良好的稳定性，相对轻质，可以创造出具有弹性的立面形式。金属网可以用于面向建筑开放空间的半透明的门帘，以及起到阻隔视线作用的幕墙、立面等。

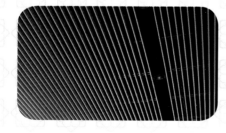

Metal Grill 金属格栅

Metal grill is usually made of steel, stainless steel or aluminum by extruding or electric welding. Crossed lattices can be designed in with different sizes. The boarder of the grill will be embedded in the framework. With flexible sizes, this kind of product is an important facade material. It is usually used in facade decoration, interior decoration and external wall facade.

该材料通常由钢、不锈钢或者铝制作，经过工业方式挤压或者电焊而成。交叉形态的正交条形格子可制作成不同的间距。格栅边缘嵌入并固定在边框上，金属格栅可制作成任何尺寸，是一种重要的立面元素。在建筑的立面装饰、室内装修以及外墙的立面上应用较多。

Metal Cladding 复合金属薄板

Cladding is usually used as floorboard or shingle with coating. This product is often achieved by extruding two metals through a die as well as pressing or rolling sheets together under high pressure. Though it is relatively expensive, it is widely used for its stability, lightweight and easy processing.

该材料常用作楼面板或者墙面板，其完成面通常覆盖有一层保护膜。复合金属薄板价格相对较贵，但是它们具有良好的稳定性并且质量较轻；同时这种材料的加工简便，工业化生产程度较高，从而可以制作出精良的、经济的立面。

Metal Sandwich Panel 金属夹芯板

Metal Sandwich Panels have different surface textures. And the installation and fixing is easy. They need little maintenance and can bear a standard 80mm-thick shingle with a span ranging from 5m to 6m. This kind of material is usually used as surface and partition wall.

该金属材料可以有不同的表面肌理，安装就位非常简便，金属夹芯板一般只需要极少的维护，它能够荷载一块80mm厚的标准墙面板，跨度可达5～6m。常用于建筑的立面以及隔墙。

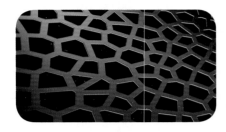

Woven Wire Mesh 金属编织物

This product is often woven with round or flat metal wire, strap or cable of stainless steel, titanium, nickel or copper. Woven metal is usually stable and can be used for large are without joints. They are great for sun-shading and screening, or used as security fencing and partition. Woven wire mesh needs little maintenance and is easy to clean. They are usually used as decorative surface.

该材料是用圆的或者扁平的金属丝、带子或索编织形成，常用材料包括不锈钢、钛、镍甚至铜。金属编织物有良好的稳定性，可用于超大的面积而中间没有接缝或者连接件。它们很适合用于遮阳和遮挡视线，也可以用于安全栅栏甚至分隔物。这种类型的材料需要很少的维护，且易于清洁，一般用于建筑表皮的装饰。

Tile Facade 砖材立面

Tiles are widely used in construction because of excellent durability and fine texture. Colorful tiles will well decorate the buildings. Moreover, they are easy to clean, fireproof, waterproof, wear-resistant and corrosion-resistant, and need little maintenance. They are often used in the areas with heavy air pollution. According to their functions, tile materials can be classified as exterior wall tiles, interior wall tiles, indoor floor tiles, outdoor floor tiles and special tiles. And according to the processing technology, tiles used for facade include six categories:

砖材因为其坚固耐用，具备很好的耐久性和质感而被广泛的运用到建筑工程当中。砖材的色彩鲜艳，具有丰富的装饰效果，并具有易清洗、防火、抗水、耐磨、耐腐蚀和维护费用低等特点。适宜于在环境污染比较大、空气灰尘多的地区。按适用场所分类，可分为外墙砖、内墙砖、室内地砖、室外地砖、特殊用砖。立面砖材按工艺及特色大致分为六类：

Glazed Tile 釉面砖

Glazed tile is a type of ceramic tile covered with a glaze to make it stain and water resistant. There are two kinds of glazed tiles: glazed ceramic tile and glazed porcelain tile. The former is made of ceramic with strong absorption capacity, weak rigidity and red back; while the later is made of porcelain with weak absorption capacity, strong rigidity and gray back. Glazed tiles can be glazed with different patterns and colors. In addition to being used for flooring, glazed tiles can be used to create accents in walls and doorways, and as trivets for hot spots.

所谓釉面砖就是砖的表面经过烧釉处理的砖，一般可分为陶制釉面砖和瓷制釉面砖两种。陶制釉面砖由陶土烧制而成，吸水率较高，强度相对较低，背面颜色为红色；瓷制釉面砖，即由瓷土烧制而成，吸水率较低，强度相对较高，背面颜色是灰白色。釉面砖表面可以做各种图案和花纹，色彩和图案比抛光砖丰富，因为表面是釉料，所以耐磨性不如抛光砖。釉面砖被广泛使用于墙面和地面装修。

Full-body Tile 通体砖

Full-body tile is made from rock fragments by high-pressure pressing. After being polished, it is hard like stone with low absorption capacity and wear resistance. They are widely used in flooring of living rooms, doorways and outdoor paths. Most of the anti-slip tiles are full-body. Full-body is designed without glaze, and its surface has the same color and material as its back. Its rough surface will not cause strong refection, thus it is widely used in the decoration of the interior and exterior spaces.

该材料是将岩石碎屑经过高压压制而成，表面抛光后坚硬度可与石材相比，吸水率更低，耐磨性好，被广泛使用于厅堂、过道和室外走道等装修项目的地面，多数的防滑砖都属于通体砖。通体砖表面不施釉，而且正面和反面的材质和色泽一致，因此得名。同时由于其表面粗糙，光线照射后产生漫反射，反光柔和不刺眼，对周边环境不会造成光污染，被广泛使用于室内的大厅、过道、墙面和室外的外墙、走道、广场等区域的装修。

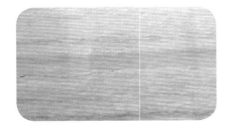

Polished Tile 抛光砖

Polished tile comes from full-body tile by polishing the surface. Compare to full-body, polished tile is smooth, hard and were resistant. They are suitable for many interior spaces like balcony and external walls except the bathroom and kitchen. With penetration technology, polished tile can be created to imitate stone or wood. The biggest characteristic is that it is harmless without any radioactive element. Moreover, there is almost no chromatic aberration for a same batch of products. Polished tiles are usually used for floor and external wall because of the lightweight and skid resistance.

该材料是通体砖坯体的表面经过打磨而成的一种光亮的砖。相对通体砖而言，抛光砖表面要光洁得多。抛光砖坚硬耐磨，适合在除洗手间、厨房以外的多数室内空间中使用，比如用于阳台，外墙装饰等。在运用渗花技术的基础上，抛光砖可以做出各种仿石、仿木效果。其最主要的特点是无放射元素，不会对人体造成伤害；同时基本可控制无色差，抛光砖经精心调配，同批产品花色一致，基本无色差。由于其砖体轻、重量轻以及防滑的特点，抛光砖经常被用作地面铺装、外墙立面等。

Vitrified Tile 玻化砖

Vitrified tile is a tile which has been processed in such a way that it has very very low porosity(and water absorption) which make it stain resistant and very strong like natural stone. Ceramic tile with a water absorption rate lower than 0.5% is called vitrified tile. And the polished vitrified tile is hard and scratch resistant with low absorption capacity and little chromatic aberration. Free of radon gas, vitrified tile bears stable physical performance, thus it is environmental friendly to be widely used in the living rooms and entrance areas instead of natural stone.

该材料作为一种常见的外墙砖，其具有天然石材的质感。吸水率低于0.5%的陶瓷都称为玻化砖，抛光砖吸水率低于0.5%也属玻化砖，将玻化砖进行镜面抛光即得玻化抛光砖，因为吸水低的缘故其硬度也相对比较高、不容易有划痕、色差小。这种产品不含氡气，各种理化性能比较稳定，符合环境保护发展的要求，是替代天然石材较好的瓷制产品，广泛用于客厅、门庭等地方。

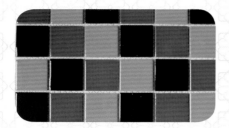

Mosaic Tile 马赛克砖

Mosaic tiles are a classification of tiles used for external walls. Usually a big mosaic tile is composed of tens of smaller pieces. They are made of various materials including ceramic, porcelain, marble and glass. Glass mosaic can be cut by machine on the single side or double sides, or it can be cut by hand. Non-glass mosaic products include ceramic mosaic, stone mosaic, metal mosaic, photoluminescent mosaic, etc. Today, mosaic tiles are popular for creating elaborate patterns like those found on floors and walls.

该材料是一种特殊的外墙砖，它一般由数十块小块的砖组成一个相对的大砖。一般分为陶瓷马赛克、大理石马赛克、玻璃马赛克。马赛克按照材质、工艺可以分为若干不同的种类，玻璃材质的马赛克按照其工艺可以分为机器单面切割、机器双面切割以及手工切割等，非玻璃材质的马赛克按照其材质可以分为陶瓷马赛克、石材马赛克、金属马赛克、夜光马赛克等等。马赛克砖主要用于墙面和地面的装饰。

Rustic Brick 仿古青砖

Rustic tile is a building material made from clay to offer a rustic, natural look. Clay comes from aluminum silicate minerals after years of weathering. When the clay is totally baken for tiles, it will turn red to be red bricks; while when it is half baken and cooled by water during this process, FeO will generate and result in black color and black bricks. This tile features influence from ancient Chinese art and architecture, along with a distressed, natural finish. Then Rustic tiles are mainly used for the renovation of historic buildings or for the facade decoration of cultural buildings or some stylish buildings.

该材料是粘土烧制的，粘土是某些铝硅酸矿物长时间风化的产物，具有极强的粘性而得名。将粘土用水调匀后制成砖坯，放在砖窑中煅烧便制成砖。粘土中含有铁，烧制过程中完全氧化时生成三氧化二铁呈红色，即最常用的红砖；而如果在烧制过程中加水冷却，使粘土中的铁不完全氧化而生成低价铁（FeO）则呈青色，即青砖。主要用于历史建筑的修复、文化建筑以及个性建筑的立面装饰等。

Stone Facade 石材立面

Natural stone is the stone extracted from natural rocks and is one of the earliest building materials in the history of mankind. Most of the natural stones have high strength, good durability and rich reserves and are easily mined, and have been favored, often used for walls, floors, roofs, building components and sculptures, etc. At first, it is largely used as structural and decorative material, but now it is no longer used as structural material but just as decorative material inside and outside of the building. Due to the natural characteristics, it always brings a deep and solemn sense of culture and history. Marble, granite, slate, sandstone and travertine are the common building stones.

天然石材是指在自然岩石中开采所得的石材，它是人类历史上应用最早的建筑材料之一。由于大部分的天然石材具有强度高，耐久性好，蕴藏量丰富，易于开采加工等特点，因此被人们所青睐，常被用作墙体、地面、屋顶、建筑构件、雕塑等材料来使用。石材在建筑中最初主要作为结构以及装饰材料出现，发展到今天，石材作为结构材料几乎已经绝迹，一般仅作为建筑内外表皮装饰材料使用。由于石材的天然特性，它经常能给人带来深厚凝重的文化感和历史感。常见的建筑石材主要有大理石、花岗石、板岩、砂岩以及凝灰石。

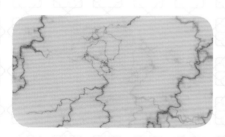

Marble 大理石

Marble is a non-foliated metamorphic rock composed of recrystallized carbonate minerals. It has exquisite, solid texture, various beautiful colors and patterns, but it could easily get weathered and has poor wearing resistance. It would gradually lose its luster after a long-term exposure in the outdoor, therefore, generally it used only for interior decoration, such as indoor walls, floors, railing, window sill, service counter and the facing of elevator room. However, pure white marble and green coloration are the exception, they are pure in quality and stable and durable enough to resist the outdoor condition.

该种石材属于变质或者沉淀的磷酸盐类岩石，其组织细密、结实，有各种美丽的颜色以及石纹。大理石纹理美观，花色繁多，品种丰富。然而易风化、耐磨性差，长期暴露在室外条件下会逐渐失去光泽，一般只用于建筑室内装饰面，例如室内墙面、地面、栏杆、窗台板、服务台、电梯间门脸的饰面等。少数的大理石如汉白玉、艾叶青等质纯、杂质少且比较稳定耐久的品种可用于室外。

Granite 花岗石

Granite is a common type of igneous rock which is granular and phaneritic in texture. This rock consists mainly of feldspar, quart and mica. It has fine texture, wear-resisting, pressureproof and anticorrosion, almost can be used for various indoor and outdoor conditions. Compared to marble, most granites are lack of peculiar stripe, just have color spot or in pure color, and color and texture are the only superiority to demonstrate their performance. Granite is largely used for wall footing, exterior wall facing, and even high-rise and super high-rise external mounting curtain wall (after technic processing).

该种石材属于火成岩，其主要成分为长石、石英、云母等，它结构细密、耐磨、耐压、耐腐，几乎可以用于室内外各种条件之下，其表现一般为均匀颗粒状以及发光云母颗粒，除少数品种以外，大部分花岗石的表现效果较为单一，与大理石相比缺少特殊的花纹，主要靠整体色彩及质感显示效果。多用于建筑的墙基础和外墙饰面，利用技术加工，可在高层和超高层外挂石幕墙。与大理石相比，花岗石石材没有彩色条纹，多数只有彩色斑点、纯色。

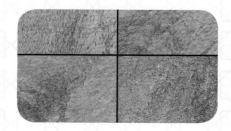

Slate 板岩

Slate is a fine-grained, foliated, homogeneous metamorphic rock derived from an original shale-type sedimentary rock composed of clay or volcanic ash through low-grade regional metamorphism. It is the finest grained foliated metamorphic rock. When expertly "cut" by striking parallel to the foliation, with a specialized tool in the quarry, many slates will form smooth flat sheets of stone which have long been used for roofing and floor tiles and other purposes. Its hardness and abradability are weaker than that of granite but stronger than that of marble. It has low water absorption and has acid proof and resistant to weathering. It is relatively cheap and a building material with high cost performance. It is largely used for roof tiles, exterior floor, interior floor and exterior walls.

该石材属于水成岩，天然板岩拥有一种特殊的层片状纹理，纹理清晰、质地细密，沿着片理不仅易于劈分，而且劈分后的石材表面显示出自然的凹凸状纹理，可制作成片状用于墙面、地面铺装等，显示出自然亲切的视觉感受。同时，板岩还经常被用作于片屋瓦。板岩的硬度和耐磨度介于花岗石和大理石之间，具有吸水率低、耐酸、不易风化等特点，板岩价格相对便宜，是一种性价比相当好的建筑用材。主要用于建筑屋面瓦、外部地板、内部地板和外墙。

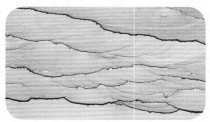

Sandstone 砂岩

Sandstone forms from beds of sand in low-lying areas on the continents with a long period of accumulation and crustal movement. Sandstone is a honed stone produces no strong reflected light and is visually soft and warm. Compared with marble and granite, sandstone has zero radiations, no harm to human body, and is suitable for large area application. It has almost the same durability as that of marble and granite, and is resistant to weathering and not easily gets tarnished. It is largely used for exterior wall, interior decoration and garden landscape.

该材料由沙粒经过水流冲蚀沉淀于河床之上，经过长时间的堆积和地壳运动而成。砂岩是一种亚光石材，不会产生强烈的反射光。视觉柔和亲切。与大理石以及花岗石相比，砂岩的放射性几乎为零，对于人体毫无伤害，适合于大面积的应用，砂岩在耐用性上可以比拟大理石、花岗石，它不易风化和变色。多用于建筑外墙、室内装饰、园林景观等。

Travertine 凝灰石

Travertine, also known tufa, often has many pores on the surface. Travertine is a terrestrial sedimentary rock, formed by the precipitation of carbonate minerals from solution in ground and surface waters or geothermally heated hot-springs. It is a quick process, and some organics and gases can't be released timely, therefor pores come into being after solidification, forming a beautiful texture. In case of stone cracks, particular attention should be paid to the connection details of the slabstone while it is used for curtain panel. Travertine has clean texture and exists in white, tan, red and cream-colored varieties. The cream-colored is the most common one (most of them are from Italy) that use for façade wall, reflecting a modern style of classic calm rather than sensationalism.

凝灰石又称石灰华、钙华，由于表面常有许多孔隙而俗称洞石，它是钙质碳酸盐在富含石灰的河流、湖泊或者池塘里快速沉淀而形成的，由于沉淀速度快，钙质碳酸盐中的一些有机物和气体不能及时释放，长久固化后便产生了孔隙，形成了美丽的纹理。在作为幕墙板材时应该特别注意石板的连接细节，防止石材裂缝的产生。凝灰石的纹理清晰，色彩有白色、灰褐色、红色、黄色等，最见的为柔和的米黄色系（多产自意大利），一般在立面幕墙上运用比较多，体现一种经典沉稳而非哗众取宠的现代风格。

Concrete Facade 混凝土立面

Generally concrete is a composite construction materials made primarily with cement (binding material), aggregate (sand, stone) and water. There are many formulations of concrete, which provide varied properties, and concrete is the most-used man-made product in the world.

混凝土是指由胶凝材料将集料胶结成整体的工程复合材料的统称。通常讲的混凝土是指用水泥作胶凝材料，砂、石作集料，与水（加或不加外加剂和掺合料）按一定比例配合，经搅拌、成型、养护而得的水泥混凝土，也称普通混凝土。混凝土是土木工程中用途最广、用量最大的一种建筑材料。

Fiber-Reinforced Concrete 纤维混凝土

Fiber-reinforced concrete (FRC) is concrete containing fibrous material which increases its structural integrity. It contains short discrete fibers that are uniformly distributed and randomly oriented. Fibers include steel fibers, glass fibers, nylon fibers, etc. Compared with ordinary concrete, FRC has high ultimate tensile and flexural strength, especially the toughness.

纤维混凝土指水泥混凝土中掺入适量散乱短纤维以提高抗裂、抗冲击等性能的复合材料。常用的纤维有钢纤维、玻璃纤维、尼龙纤维等。与普通混凝土相比，纤维混凝土具有较高的抗拉与抗弯极限强度，尤以韧性提高的幅度为大。

Aerated Concrete Block 加气混凝土砌块

Aerated concrete block is a new building material that is lightweight, polyporous, can be nailed, sawn and planed, and has high performance of heat insulation, fire resistance and certain shock resistance. In addition, it is environmentally friendly and largely used for high and low rise buildings.

加气混凝土砌块是一种轻质多孔、保温隔热、防火性能良好、可钉、可锯、可刨和具有一定抗震能力的新型建筑材料。这种优良的新型建筑材料还具有环保等的优点，被广泛使用于高低层建筑中。

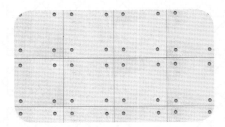

Fair-Faced Concrete 清水混凝土

The fair-faced concrete is known as decorative concrete for its great decorative effect. It is defined as a kind of one-forming concrete without any additional ornament, but serving its function as a natural and solemn facing of the building itself. Fair-faced concrete has smooth surface, uniform color, clear edges and corners, free of clashing and pollution, just need to be coated with one or two layers of transparent protectants. It can be widely used for ancient buildings, municipal bridges, sports venues, highway guardrails and external wall base of large factory.

清水混凝土又称装饰混凝土，因其极具装饰效果而得名。它属于一次浇注成型，不做任何外装饰，直接采用现浇混凝土的自然表面效果作为饰面。它表面平整光滑、色泽均匀、棱角分明、无碰损和污染，只是在表面涂一层或两层透明的保护剂，显得十分天然、庄重。可广泛应用于古建筑物、市政桥梁、运动场馆、高速公路的护栏、大厂房等外墙基面涂装。

Carbon Cast Steel 碳素铸钢

Carbon cast steel is a precast concrete that reinforce carbon fiber grilling to cope with the reinforcement force and shear transfer during construction. Due to carbon fiber reinforcement is corrosion resisting, it doesn't require too much concrete to cover the carbon steel precast concrete products. Compared to the traditional precast concrete and competitive construction methods, its durability has been improved, and even lighter and stronger than before.

碳素铸钢是一种用碳纤维格栅加固以应对施工过程中的加固力或剪力传递的预制混凝土。由于碳纤维加固结构抵抗腐蚀，因此碳素铸钢预制混凝土产品不需要过多的混凝土覆盖，与传统的预制混凝土和有竞争力的建筑方法相比耐久度有所提高，重量更轻且持久力更强。

Translucent Concrete 透明混凝土

Translucent concrete is made of ordinary concrete and fiber with light-transmissive properties. Its stiffness is greatly enhanced by a special chemical additive and it is as strong as traditional concrete. At night, when the light transmitted through various translucent concrete, a dreamy color world has been created. In the daytime, abundant natural light may reduce the use of indoor light, thus saving energy.

透明混凝土由普通混凝土和玻璃纤维组成，这种新型混凝土可透过光线。透明混凝土跟传统混凝土一样坚固，原因是一种特殊的化学添加剂大大增加了它的坚硬度。夜晚，光线透过不同玻璃质地的透明混凝土，营造出梦幻的色彩效果，而自然光的射入也可以减少室内灯光的使用，从而节约能源。

Glass Facade 玻璃立面

The function of glass as construction material is more than lighting; it is required for lighting adjustment, insulation, security (bullet-proof, burglarproof, fireproof, radiation-resistant, electromagnetic proofing), art deco etc. The types and processing way of glass have obtained new development with the continuous development of demands. New types i.e. laminated glass, Ion exchange glass, ceramic enameled glass, chemical thermal decomposition glass, cathode sputtering glass etc lead to rapid increase of construction consumption. Glass has become the third largest construction material after concrete and steels.

玻璃作为建筑材料，其功能不再仅仅是满足采光要求，而是要具有能调节光线、保温隔热、安全（防弹、防盗、防火、防辐射、防电磁波干扰）、艺术装饰等特性。随着需求的不断发展，玻璃的成型和加工工艺方法也有了新的发展。现在，已开发出了夹层、离子交换、釉面装饰、化学热分解及阴极溅射等新型玻璃，使玻璃在建筑中的用量迅速增加，成为继水泥和钢材之后的第三大建筑材料。

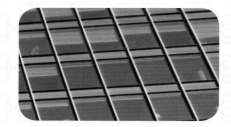

Low-E Glass Low-E 玻璃

LOW-E, Low Emissivity for short, is a energy-saving glass opposed to heat reflective glass. Its coating layer with high transmission of visible light and high reflection of far infrared brings superior insulation and transmission effect compared with ordinary glass and traditional coated glass for construction. It applies to architectures of transparent, highly permeable externals and natural lighting, effectively refraining from light pollution.

LOW-E 为英文 Low emissivity 的简称，又称低辐射镀膜玻璃，是相对热反射玻璃而言的，是一种节能玻璃。其镀膜层具有对可见光高透过及对中远红外线高反射的特性，使其与普通玻璃及传统的建筑用镀膜玻璃相比，具有优异的隔热效果和良好的透光性。适用于外观设计透明、高通透性，自然采光的建筑物，可有效避免"光污染"。

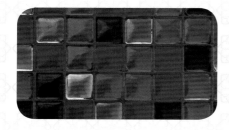

Ceramic Enameled Glass 釉面玻璃

Ceramic enameled glass refers to glass with pretty colors or patterns made from cutting glass with a layer of colorful soluble glaze on the surface being sintered, annealed or steeled to firmly agglutinate the glaze layer and glass. It applies to external wall finishes of architectures for its superior chemical stability and decorativeness.

釉面玻璃是指在一定尺寸切裁好的玻璃表面上涂敷一层彩色的易溶釉料，经烧结、退火或钢化等处理工艺，使釉层与玻璃牢固结合，制成的具有美丽的色彩或图案的玻璃材料。它具有良好的化学稳定性和装饰性，适用于建筑物内外墙饰面。

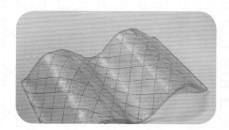

Wired Glass 夹丝玻璃

Wired glass, alias shatterproof glass is ordinary slate glass heated to be glowing red and soften, with preheating wires or wire netting being pressed into the middle. It's superior in fire proofing that it could shelter from flames and it will not crack in high temperature combustion; it will not cause injury by fragments. Additionally it is burglarproof that wire net is another shield behind the glass. It applies to roof and balcony windows etc.

夹丝玻璃又称防碎玻璃，是将普通平板玻璃加热到红热软化状态时，再将预热处理过的铁丝或铁丝网压入玻璃中间而制成的玻璃。它的特点是防火性优越，可遮挡火焰，高温燃烧时不会炸裂，破碎时不会造成碎片伤人。另外还有防盗性能，玻璃割破还有铁丝网阻挡。适用于屋顶天窗、阳台窗等。

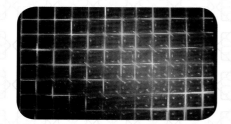

Laser Glass 镭射玻璃

Laser glass, alias holographic glass or laser holographic glass, is an innovative decorative glass product developed with latest holographic technology. It is a type of laminated glass that prefabricated laser holographic film is placed between two pieces of glass with laser holographic film technology to make a transparent surface but appearing colorful in different angles under the light refraction. It applies to architectures like bar, hotel and cinema etc.

镭射玻璃亦称全息玻璃或镭射全息玻璃，是一种应用最新全息技术开发而成的创新装饰玻璃产品。镭射玻璃是一款夹层玻璃，应用镭射全息膜技术把预制镭射全息膜夹在两层玻璃中间，形成表面透明，但在光线的折射下，从各种不同角度上看可呈现不同的颜色。镭射玻璃常用于酒吧、酒店、电影院等建筑中。

Insulating Glass 中空玻璃

Insulating glass is a new construction material with superior heat and sound insulation functions, beauty and applicability, and reducing weights of buildings. It is composed of two or three pieces of glass adhesive to aluminum flame with drier inside, by composite binder of high intensity and air seal. It mainly applies to architectures which require heating, air conditioning, noise proof, no direct sunlight and special light.

中空玻璃是一种良好的隔热、隔音、美观适用、并可降低建筑物自重的新型建筑材料。它是用两片(或三片)玻璃,使用高强度高气密性复合粘结剂,将玻璃片与内含干燥剂的铝合金框架粘结,制成的高效能隔音隔热玻璃。中空玻璃主要用于需要采暖、空调、防止噪音以及需要无直射阳光和特殊光的建筑物上。

Coating Facade 涂料立面

Coating is a generic terms of materials converted to a solid protective film after application to a surface. Coating facade mainly refers to a wall painted with coating. Architectural coating functions as a layer of decoration and protection for architectures, enhancing the service durability of main building materials. It applies generally to residences, industrial buildings, commercial buildings and public buildings etc.

涂料是指应用于物体表面而能结成坚韧保护膜的物料的总称。涂料立面主要是指用涂料进行装饰涂敷的建筑墙面。建筑涂料起到装饰和保护建筑物的作用,能够提高主体建筑材料的耐久性。建筑涂料通常适用于住宅、工业建筑、商用建筑、公共建筑等。

Acrylic Exterior Coating 丙烯酸系外墙涂料

It is a solvent-based exterior coating made by grinding modified acrylic copolymer as the film-forming material, with ultraviolet absorbent, filler, organic solvent, accessory ingredient etc. It is one of the most favorite exterior coatings at present for its performances of color maintenance, decorative effect and long service durability.

以改性丙烯酸共聚物为成膜物质,掺入紫外光吸收剂、填料、有机溶剂、助剂等,经研磨而制成的溶剂型外墙涂料。具有保持原色、装饰效果好、使用寿命长等特点,是目前外墙涂料中较为常用的涂料之一。

Polyurethane Exterior Coating 聚氨酯系外墙涂料

It is a high-quality exterior coating made from polyurethane resin or composite of polyurethane and other resin as main film-forming materials, with filler and accessory ingredient etc. It is characterized by high elasticity, flavor decorative effect, able to sustain severe stretching and decoration effect up to 10 years. It applies to the decoration on exterior walls of concrete and cement mortar, i.e. the exterior wall of upscale residences and hotels etc.

以聚氨酯树脂或聚氨酯与其它树脂复合物为主要成膜物质,加入填料、助剂组成的优质外墙涂料。其特点为涂料弹性高、装饰性好,可以承受严重拉伸,装饰效果可达 10 年。适用于混凝土或水泥砂浆外墙的装饰,如高级住宅、宾馆等建筑物的外墙面。

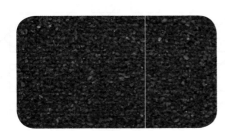

Color Sand Coating 彩砂涂料

Color sand coating is made from high-temperature clinkering color ceramic granules or naturally colored stone chips as aggregate with acrylic copolymer emulsion as adhesive and various accessory ingredients like additive etc. It is nontoxic, quick drying, glare-resistant, non-fading with high pollution-resistant performance and no solvent pollution. Different arrangement of aggregate will lead to varied levels of deep colors, obtaining textures of rich colors resembling natural stone. It mainly applies to the decoration of exterior walls with planking and cement mortar surfaces.

彩砂涂料是以丙烯酸共聚乳液为胶粘剂,由高温燃结的彩色陶瓷粒或以天然带色的石屑作为骨料,外加添加剂等多种助剂配置而成。该涂料无毒,无溶剂污染,快干,耐强光,不褪色,耐污染性能好。利用骨料的不同组配可以使深层色彩形成不同层次,取得类似天然石材的丰富色彩的质感。彩砂涂料主要用于各种板材及水泥砂浆抹面的外墙面装饰。

Inorganic Coating 无机涂料

Inorganic coating used in exterior walls is made after mixing, blending, grinding with potassium silicate or silica sol as main adhesive and filler, paint and other accessory ingredients. It is resistant to ageing and ultraviolet radiation, easy for construction with film forming in low temperature, generally used in interior and exterior wall decoration of industrial and civil buildings.

外墙所用的无机涂料是以硅酸钾或硅溶胶为主要胶结剂,加入填料、颜料及其它助剂等,经混合、搅拌、研磨制成的。具有成膜温度低、耐老化、抗紫外线辐射、施工方便等优点。无机涂料常用于工业与民用建筑物的外墙和内墙饰面材料。

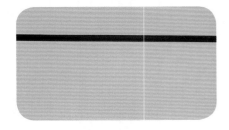

Fluorocarbon Coating 氟碳涂料

Fluorocarbon coating takes fluorine resin as the main film forming material. It occupies excellent performances i.e. weather resistance, thermo tolerance, low temperature resistance, chemical resistance and unique non-adhesion and low friction, owing to the great electro negativity of fluorine and the strong carbon-fluorine bond. Its superior comprehensive performances and price/performance ratio brings it advantage beyond comparison on super high-rise buildings, landmark buildings and key projects etc.

氟碳涂料是指以氟树脂为主要成膜物质的涂料,由于引入的氟元素电负性大,碳氟键能强,具有特别优越的耐候性、耐热性、耐低温性、耐化学药品性,而且具有独特的不粘性和低磨擦性。氟碳涂料优异的综合性能和性能价格比,使它在超高层建筑、标志性建筑、重点工程等方面具有无以伦比的竞争优势。

Wooden Facade 木质立面

The wooden material is superior for it can give people a warm and natural feeling both from the visual effect and tactile impression, but the natural wooden materials have poor weather resistance and durability, easy to be damaged and flammable, and its repair and maintenance are also quite complex, so that it rarely uses natural untreated wood as the building facade. Generally speaking, it will use anti-corrosion technology or other processing methods to make high-quality durable wooden materials for building facades.

木材的优点无论是观感还是触感均给人以亲切自然的感受，然而天然木材耐候性差，耐久年限较少，易损伤，易燃，维修保养较为复杂，这就决定了很少直接选用未经处理的天然木材来做建筑的外立面。一般采用防腐工艺或其他加工方法，制成经久耐用的高品质木质板材用于建筑的外立面。

Wooden Mosaic 木质马赛克

Wood mosaic is the outer and inner wall decoration materials, using wood and bamboo as the raw material and being processed by manual or mechanical methods to form glazed-tile-like mosaic appearance. Wood mosaic has various styles with advantages of durability, heat insulation, sound insulation and no accumulation of dust, usually used in the ground, facade, ceiling and other walls.

木质马赛克是属于外内墙装修装饰材料类，是以木、竹材为原料，用手工或机械的方法加工出外形如釉面砖的马赛克。木质马赛克品种、式样繁多，经久耐用、保温、隔音、不积尘埃，通常用在地面、立面、天棚以及其它需要装修装饰的墙面上。

Heavy Bamboo 重竹

Heavy bamboo is a kind of bamboo being reorganized and strengthened to form another new bamboo material. The density of heavy bamboo material is 1.5 times higher than the bamboo before being processed. Its hardness is neck-and-neck with the high-grade hardwood, not rotten, not cracking and not deformed. It substitutes for stone, steel, concrete and other materials to be widely used in outdoor, bathrooms and other damp and odious environment with big difference in temperature.

重竹是将竹材重新组织并加以强化成型的一种新型竹质材料。重竹材料的密度是加工前竹子的1.5倍，硬度与高档硬木相当，而且久泡不朽、久晒不开裂不变形。可替代石材、钢铁、混凝土等材料广泛应用于户外、卫生间等潮湿及温差大的恶劣环境。

Artificial Board 人造板

Artificial board uses wood or other non-wood plants as the raw materials to be processed into component materials such as veneer, particles or fibers in various shapes, then applies (or not) adhesives and other additives to form the new composite board materials. The artificial board is huge, well structured and convenient; it is not easily expanded or shrunk with stable size, well-distributed material quality, and not easy to deform or crack.

人造板是以木材或其他非木材植物为原料，加工成单板、刨花或纤维等形状各异的组元材料，经施加（或不加）胶黏剂和其他添加剂，重新组合制成的板材。人造板幅面大，结构性好，施工方便；膨胀收缩率低，尺寸稳定，材质较锯材均匀，不易变形开裂。

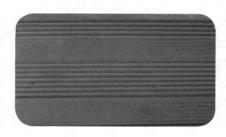

WPC Board WPC 板

WPC board uses wood as the main raw material in proper treatments; after acting with various plastic in different methods, it generates a composite material of high performance and high additional value. WPC board is based on the high density polyethylene and wood fiber, so it has the characteristics of water resistance, corrosion resistance, and long service life, etc. It is widely and mainly used for building, furniture and other industries.

WPC 板是以木材为主要原料，经过适当的处理使其与各种塑料通过不同的复合方法生成的高性能、高附加值的新型复合材料。WPC 板的基础为高密度聚乙烯和木质纤维，具有耐水、耐腐性能、使用寿命长等特点。广泛用于建材、家具等行业。

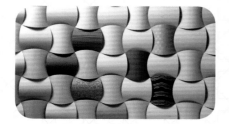

Wovin Wall Wovin 立体瓦编墙

Wovin Wall is a kind of innovative building material, only using 2 tile types and 5 kinds of material changes; when being used in an ordinary space, whether as a theme or as a background, it can create the feeling of everything fresh and new. The users can print thousands of colors and self-design image combination in the tiles, or through simple assembling, they can swap tiles themselves, thus creating a changeable wall.

Wovin 立体瓦编墙是一种创新建材，仅利用2种瓦片型式，再搭配5种材质的变化，便能在平凡的空间里，不论是作为主题或作为背景，都能营造出耳目一新的感觉。使用者甚至可以在瓦片上喷印千百种颜色与自定的图像组合，通过简易的组装方式，DIY 抽换瓦片，就能创造出多变的墙面姿态。

Plastic Facade 塑料立面

Under the conditions of the existing technology, the building facades mainly use transparent plastic, such as poly methyl methacrylate (PMMA), polycarbonate (PC), tetrafluoroethylene (ETFE), glass fiber reinforced polymer (GRP), polyvinyl chloride (PVC) and polystyrene (PS), etc. The polycarbonate with more mature technology is used more often in the building facade and roof than any other plastic materials.

在现有的技术条件下，建筑外立面运用的塑料以具有一定特点的透明塑料居多，如聚甲基丙烯酸甲脂（PMMA），聚碳酸酯（PC），四氟乙烯（ETFE），玻璃纤维增强树脂（GRP），聚氯乙烯（PVC），聚苯乙烯（PS）等，其中以聚碳酸酯在建筑外立面及屋面中应用最多，技术也相对成熟。

Polycarbonate (PC) Board 聚碳酸酯（PC）板

The main components of polycarbonate (PC) board is polycarbonate, which is a high quality engineering plastic plates produced with CO-EXTRUSION technology. Because its surface is covered with a high-concentration ultraviolet absorbent, it bears the characteristics of UV resistant, long-term weather resistance and non-fading. It is a high-tech, energy-saving and environment-friendly plastic board with excellent comprehensive performance, a plastic building material widely used in the international.

聚碳酸酯（PC）板是以聚碳酸酯为主要成分，采用共挤压技术 CO-EXTRUSION 而成的一种高品质的工程塑料板材。由于其表面覆盖了一层高浓度紫外线吸收剂，除具有抗紫外线的特性外并可保持长久耐候、永不褪色的特点。是一种高科技、综合性能极其卓越、节能环保型塑料板材，是目前国际上普遍采用的塑料建筑材料。

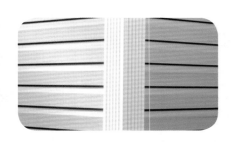

PVC Board   PVC 板

PVC board keeps PVC as the raw material to form a material with cross section in honeycomb mesh structure. It is produced under the interaction of polyvinyl chloride resin and stabilizer, bearing the advantages of excellent corrosion resistance, insulativity, heat resistance and impact resistance. The surface of the board is smooth without bubbles or cracks and can be colored based on the users' needs. It is widely used, especially in the building materials industry which 60% of the materials adopt PVC laminates.

PVC 板是以 PVC 为原料制成的截面为蜂巢状网眼结构的板材。该产品系聚氯乙烯树脂与稳定剂等辅料配合后压延，层压而成，具有优质的防腐蚀性、绝缘性、耐温性和耐冲击性。外观板面光滑平整，无汽泡，无裂缝，可以根据用户需要配色。其应用范围较广，在建材行业中占的比重最大，为 60%PVC 层压板。

Acrylic (PMMA)   亚克力（PMMA）

Its chemical name is polymethyl methacrylate, commonly known as perspex; this is one kind of early important thermoplastic plastic, with superior transparency, chemical stability and weather resistance, easily colored, easily processed and beautiful appearance, and it is widely used in the architecture industry. The products usually include casting plate, extrusion board and moulding compound, and the boards can be processed into the surface plate, corrugated plate and perforated plate, etc.

亚克力（PMMA）化学名称为聚甲基丙烯酸甲酯，俗称有机玻璃，是一种开发较早的重要热塑性塑料，具有较好的透明性、化学稳定性和耐候性，易染色，易加工，外观优美，在建筑业中有着广泛的应用。产品通常可以分为浇注板、挤出板和模塑料，板材可以加工为平板、波纹板、穿孔板等。

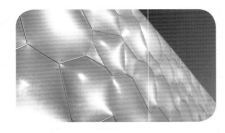

ETFE Film   ETFE 薄膜

ETFE is the toughest fluoroplastic; it preserves the good heat resistance, chemical resistance and electrical insulation performance from PTFE, at the same time, its radiation resistance and mechanical properties have been greatly improved and the tensile strength can reach 50 MPa, nearly 2 times of that of PTFE. ETFE film is mainly used as the roof membrane material of various special-shaped buildings, such as the stadium bleachers, cone top of the building, casino, revolving restaurant awning, entertainment hall roofing cover, parking lot, galleries and museums, etc.

ETFE 是最强韧的氟塑料，它在保持了 PTFE 良好的耐热、耐化学性能和电绝缘性能的同时，耐辐射和机械性能有很大程度的改善，拉伸强度可达到 50MPa，接近聚四氟乙烯的 2 倍。ETFE 薄膜主要作为各种异型建筑物的篷膜材料，如运动场看台、建筑锥型顶、娱乐场、旋转餐厅篷盖、娱乐厅篷盖、停车场、展览馆和博物馆等。

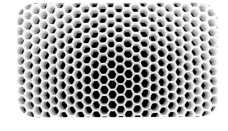

Plastic Honeycomb Board   塑料蜂窝板

Plastic honeycomb board is composed of two pieces of thin panel, firmly bonding on both sides of a layer of thick honeycomb core material, also known as the honeycomb sandwich structure. It has light quality, sound insulation effect and good elasticity, so it is commonly used as one of the building materials. Generally, it is widely used in decoration, curtain wall, roof and floor, etc.

塑料蜂窝板是由两块较薄的面板，牢固地粘结在一层较厚的蜂窝状芯材的两面而制成的板材，亦称蜂窝夹层结构。质轻、有隔音效果、弹性好，是常用建筑材料的一种。在装饰、幕墙、屋顶、楼板等建筑领域广泛应用。

EPS Board   模塑聚苯乙烯泡沫塑料（EPS）板

EPS Board is composed of many closed polyhedral honeycombs. Each comb has a diameter of 0.2 ~ 0.5 mm, and the comb wall is 0.001 mm think. There is about 2% polystyrene, and the rest is air. Static air inside the comb is a poor conductor of heat, so this kind of material has good heat preservation performance. With its unique characteristics of vapor permeability, high compressive strength, convenient installation and long service life, etc., it becomes an important building energy-saving insulation materials, widely used in architectural engineering.

模塑聚苯乙烯泡沫塑料（EPS）板是由很多封闭的多面体蜂窝组成的。每个蜂窝的直径为 0.2 ~ 0.5mm，蜂窝壁厚为 0.001mm。其中聚苯乙烯约 2%，其余为空气。蜂窝内的静止空气为热的不良导体，因而这种材料具有好的保温性能。加上其独特的抗蒸汽渗透性、较高的抗压强度、便捷的施工安装及长久的使用寿命等特点，成为重要的建筑节能保温隔热材料，在建筑工程中得到广泛的应用。

Contents 目录

Shopping Mall 大型购物中心

Interspar Fürstenfeld Interspar Fürstenfeld 超市　026

The architecture in no standard shape in general is a great solution to the connection between the retail area and the others, through a green and open design.

整体采用非标准对称的建筑形式，很好地解决了零售区与其他区域之间的联系，设计绿色、开放。

Pendorya Shopping Center　PENDORYA 购物中心　032

The facade uses precast concrete panels and glass, while its interior design and form are derived from the structure's particular location.

立面采用预制混凝土板与玻璃的混搭模式，内部空间根据建筑结构的特定位置打造活力型购物空间。

Metropolis　大都会　040

The design of Metropolis mimics an open-air street scene through various architecture details to create a pop, sophisticated and dynamic shopping space.

设计模拟出一个露天的街道场景，通过一系列的建筑结构细部的表现，营造出一个时尚、高端、动态的商业空间。

Dot Envelope　Dot Envelope 商场　048

The skin uses metal and concrete with a series of irregular symmetrical dot decoration to create a great visual aesthetic appearance.

建筑表皮采用金属与混凝土相结合的构造，通过一系列不规则对称的圆点装饰设计，打造了一个极具视觉美感的商场外观。

Mecenatpolis　056

Connecting to the residential and office buildings, the canyon-style mall also provides a dynamically integrated pedestrian place.

峡谷造型的商业空间作为纽带，既连接着住宅、办公空间，也营造了一个流动式的步行体验。

Skiterminal – Schladming　　施拉德明滑雪度假综合体　　064

The complex is an oversized loop composed of three independent but connected buildings, with a unique mixed facade of glass and metal materials.

整个综合体外部表现为一个环形结构，三个主要的建筑群既相互独立又互为整体，玻璃与金属材质的混搭表皮显得新颖而独特。

Herma Parking Building　　赫尔玛停车大楼　　070

The Herma parking building maximized the potential commercial uses of the land, maximized of lawful commercial area, the possibility of using front terrace, Increased the architectural value through front-skin design.

赫尔玛停车大楼的设计最大化地发挥了商业用地的潜力，最大限度的优化了商业空间，并通过建筑表皮的设计提升了建筑价值。

Office Building 办公建筑

Steel Band　　Steel Band 办公楼　　088

The steel framework, acting as a filter for the users, provides an ever changing and powerful experience when viewed from "la promenade architecturale" all around Steelband.

钢铁框架的设计保证了一定的私密性，并带来一种不断变化的强劲的视觉体验。

Aalta　　Aalta 办公楼　　096

The facade has a dynamic urban aspect: undulating through streams, its white shell seems to lead a movement in the neighborhood.

外立面采用白色波状设计，线条流畅，给整个街区带来了动感与活力。

Ahlström Salmisaari Office Building　　Ahlström Salmisaari 办公楼　　104

The exterior is dominated by red brick, both in surfaces and as a tectonic structure like the old industrial buildings.

大楼外部材料主要采用红砖及玻璃，使得建筑表面和整体结构看起来更像一个工业建筑。

Westraven　　Westraven 办公楼　　112

The ensemble has a great variety of sustainability-related aspects, such as the second-skin façade and climate ceilings as well as the underground warmth and cold storage. A series of integrated design solutions that are not only aesthetic and functional, but are useful in terms of material saving as well.

整座建筑从多方面体现了可持续的设计理念，如表皮涂层和气候调节天花板，地下储热和降温系统等，不仅美观实用，也有效地节省了材料和能源。

Head office of Quebecor 魁北克总部　　134

The materiality of the new building is in continuity with the existing building, endowing the grouping with homogeneity of color and texture, thus create a neutral-coloured link between neighboring buildings with their variety of materials.

建筑材料的选择与旧建筑一致，颜色与纹理也体现了一致性，从而建立起与临近建筑一致的色彩联系。

Manitoba Hydro Place 马尼托巴水电局大楼　　144

The building relies on passive free energy, such as materials, orientation and other characteristics provide most energy for cooling and heating, thus avoid affected by extreme climate.

该建筑主要通过被动方式来节能，例如材料、朝向等特点，为建筑制冷和采暖提供大部分能源，使建筑远离了极端的气候反应。

Centennial Place 世纪广场　　152

The most important aspect of Centennial Place is its ability to connect with the city on multiple levels. A two storey high pedestrian concourse runs between the buildings, linking to Calgary's unique elevated walkway.

该项目最显著的特点在于其在多个层面连接城市的能力，两层楼高的步行街穿过建筑，将卡尔加里市内独特的高架行人道连接起来。

Exaltis Office Tower Exaltis 办公大楼　　160

The building is a rectangular prism modified and sculpted by the introduction of two curves along its short facades. These curving surfaces transform the rigid rectangles into a fluid forms.

大楼整体为长方体造型，两个侧面稍有一些弧度，捎带弯曲的表面让刚硬的矩形外观更富动态感。

Tempo Office Tower 节奏办公大楼　　168

The tower is shaped by a skin that curves at the corners which shaped in segments, with each floor stepping incrementally in or out from the next and creating a ripple effect.

楼体的四个表面之间为曲线造型，其曲线弧度随楼层逐步递进，创造出一种水中涟漪的景象。

O-14 O-14 商业大厦　　174

With porous concrete outer frame structure, the design meets the building's ventilation, environmental protection and energy saving conditions and creates a good visual effect at the same time.

建筑外立面设计采用多孔混凝土外框架结构，形成良好视觉效果的同时满足建筑的通风透气性，环保节能。

Drexel Parking/Office Building 德雷克塞尔办公楼及停车场　　188

The structure of the office building and that of the parking facility are separate so as to eliminate the transfer of vibration or noise.

设计将办公建筑与停车场的结构分开设置，以消除车流来带的振动和噪声。

Zuellig Building　裕利大厦　194

The fan-shaped floor plates' taper towards the southern face – which allows the building's facades to be set orthogonally to the east and west property lines while also directly engaging the fabric of the city in those directions.

大厦整体建筑外型设计因地制宜，楼体呈扇形，锥尖朝南，这样布局使得楼体外墙呈正东、正西方向，更好地与城市整体结构融为一体。

Fukoku Tower　大阪富国大厦　198

Taking inspiration from the profile of a gigantic tree whose roots proliferate on the surface of the ground, splayed at its base, the tower's outline tapers elegantly as it rises, gracing the city's skyline with a vertical asymptote.

设计灵感源于一颗巨树，基座部分的镶镜立面为扩散状，与上部光滑的玻璃立面形成鲜明对比，极具标识性。

Hotel Building 酒店建筑

ME Barcelona Hotel　西班牙巴塞罗那 ME 酒店　206

The tower is composed of two volumes stuck together. The way these boxes are placed against each other creates a brand new city landmark.

大楼由两个盒子状体量"粘"在一起，这些体量互相叠放的设计，创建出一处崭新的城市地标。

NH-Fieramilano Hotel　米兰国际展览中心 NH 酒店　218

Each tower is inclined at an angle of 5 degrees and being connected by two large semi-transparent arms, form a simple yet powerful landmark.

两座酒店均有 5 度倾斜角，以半透明的悬臂结构连接，形成简单却强势的地标形象。

Hotel Well　Well 酒店　226

The selection of materials (lightweight aluminum composite panels and precast concrete panels) for the facade envelope marks the theme of duality (light / heavy); this intervention aims to establish new and more complex spatial relations of the existing space.

立面上金色的轻质复合铝板，与预制混凝土墙板形成"轻"与"重"的对比；整个建筑的设计使新建筑与原建筑形成了新的更复杂的空间对话。

Hotel in Mengibar　门希瓦尔酒店　234

The architects set a powerful front towards the road and releases all existing garden area and relaxation area and expansion of the hotel.

设计师为该建筑设计了强有力的正面，使酒店与周边的园林区、休闲区和谐相融。

The Sheraton Milan Malpensa Airport Hotel & Conference Centre
米兰马尔彭萨机场喜来登酒店及会议中心   240

Pultruded fibreglass panels on the facade has a series of qualities highly suitable for building; the tension created between solid and void, curved and straight line, the play of light reflected off and through the building, provide a variation of views and give a dynamic to the architecture.

建筑外立面采用拉挤纤维玻璃板，性能优越；曲线立面上不规则的虚实设计，配合灯光效果，产生一种深度感和动态感。

Mint Hotel Amsterdam   阿姆斯特丹 Mint 酒店   252

The facades consist of moveable perforated shutters, glazing and brickwork, emulating the spirit of traditional Dutch architecture.

建筑外立面由可移动的穿孔百叶窗、玻璃和砌砖构成，沿袭了传统的荷兰建筑精神。

Transportation and Sports Building
交通体育建筑

CTA Morgan Street Station   CTA 摩根车站   264

To meet the requirements of sustainable development, the design took the neighborhood environment as reference to choose materials or layout and made itself the golden model of the modern urban rail transit system in Chicago.

设计从可持续发展的要求出发，无论是材料的选择还是车站的布局，均以车站周边的环境作为参照物，堪称芝加哥现代城市轨道交通体系的典范。

ÖBB Rail Service Center OBB   铁路服务中心   270

The project is a sustainable "slim building" based on the terrain characteristics, with various types of materials in means of technology.

设计很好的把握地形的特征，运用多种材料和技术手段，建造了一个可持续发展的"苗条建筑"。

FK Austria Wien Training Academy   奥地利维也纳 FK 足球训练学校   276

The training hall as the main body building lower half below ground, the unique design of roof and the rational utilization of terraces are leading efficiently to the interaction between the training hall and the grass pitches.

本案的主体训练大厅位于半下沉的地面，通过屋顶的特殊设计以及建筑露台的合理运用，有效的解决了大厅与训练场的互动。

Olympic Shooting Venue   伦敦奥运射击场   284

With the concept of sustainability, it applies modular steel components and curved membrane façade to create a functional and colorful Olympic venue.

设计以可持续建筑为目标，采用模块化的钢筋结构与卷曲的薄膜装置，打造了一个功能突出、颜色鲜明的特色奥运场馆。

Four Sport Scenarios   四个体育场   292

The steel structure and geometrical roofing system with parallel strips of the main building bodies have created an open sports space, whose perforated façades lead to a single and perpendicular bioclimatic diagram.

体育场主体采用钢筋结构，屋面通过几何平行条状设计，营造了一个开放性的运动空间，多孔立面的形态勾勒出一个垂直生物气候图。

The Sports Park Stozice Stozicet   体育公园   302

The football stadium is 'sunken' in the park – only the roof over the stands rises above the plane of the park as a monolithic crater reminiscent of the pit that once formed part of the landscape; A canopy encircles the hall mirrors the scalloped shell.

足球场的表皮采用绿色屋顶与金属材质的混搭，形态如弹坑一般惟妙惟肖；体育馆则采用玻璃幕墙与铝板材质，突出扇形贝壳的衍生结构。

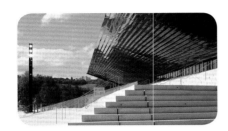

Sports Palace   皇家体育场   314

The design emphasizes the modern sense of the contours of the stadium. Outward metal façade offers a strong visual impact. The base of the steps is novel and unique.

设计强调体育场的现代轮廓感，金属表皮的外立面通过不断向外延伸的体量，形成了具有视觉冲击力的外部形态，底座台阶的设计新颖而独特。

Mall of Asia Arena   亚洲商城体育馆   322

With semi-elliptical shape structure, aluminum composite panel material for the top and glass facade, the building has a functionally-purposed three-dimensional layout.

外形上呈半椭圆形的建筑结构，顶部表皮采用铝复合板，立面运用玻璃作为主要的材料，整体采用立体式的布局，功能明确。

Richmond Olympic Oval   列治文奥运速度滑冰馆   328

Using wood, glass and steel, the build has a steel and glass mixed outlook and a curved roof.

本案在设计时将顶部设置为曲面的屋顶，在整体结构上很好的运用了木材、玻璃和钢材，外立面突出钢与玻璃的混搭装饰。

Olympic Tennis Centre, Madrid, Spain   西班牙马德里奥林匹克网球中心   334

Inside the "magic box" the tennis arenas are adapted to the different uses of the complex. The roofs of the three indoor/outdoor courts are giant mobile slabs mounted on hydraulic jacks, which interplays with the light skin.

建筑总体呈盒子形状，设计时将顶部设置为开放式的结构以满足多种用途，上部巨型移动平板与立面可移动的轻质表皮相得益彰。

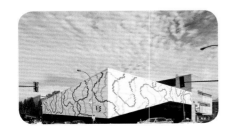

Expansion of Centre Sportif J.C. Malépart J.C. Malépart   体育中心扩建项目   348

Both indoors and outdoors, a wave develops on the perimeter of the building, rising and falling in two continuous undulations. Glass façade, milky tone and reflective quality of the upper wall are echoed in the material for the roof to ensure the extension from one to the other in contrast with the clear demarcations of the building erected fourteen years ago.

在造型设计上，室内外均采用波浪形态，玻璃立面、不透明牛奶色的墙体与屋顶材料色调相呼应，建筑整体简洁现代而具有可辨性。

Healthcare Building 医疗建筑

Children's Hospital Colorado　　　科罗拉多州儿童医院　360

Taking full advantage of natural light, the designer creates a childlike rehabilitation center for pediatric patients through large glass facade with a series of colorful decorates.

设计充分的利用自然光线，通过大面积的玻璃立面与一系列色彩艳丽的空间装饰，为儿童患者创造一个充满童趣的康复中心。

Laboratories, Haukeland University Hospital Haukeland　大学医院实验室大楼　366

The extensive use of glass in the roof, facades and interior walls have minimized the adverse effects caused by the surroudning building block as well as maximized the views and openness.

屋顶、立面和内墙的设计采用了大量的玻璃，既克服了周边建筑遮挡的不利因素，又营造了开放性的视觉空间。

Diagonal Clinic　　西班牙 Diagonal 诊所　374

The main design idea is to create a dialogue between the new building and the surroundings, then conditioned the project by the relation of the different important elements and solved the concrete program.

项目设计的主要理念是通过不同重要因素关系进行项目规划，提出解决方案，建立新建筑与周边的对话。

Kochi Health Sciences Center　　高知健康科学中心　382

Use glass, light metal and ceramic title as skin's main materials. And the cross-shaped upper floor not only enriched the building form but also create favorable landscape view.

建筑采用玻璃、轻质金属以及瓷砖等作为表皮的主要构造，十字形的上层楼面造型既丰富了建筑形式，又创造了良好的景观视线。

Art Building 艺术建筑

Alésia Archaeological Museum　　阿莱西亚考古博物馆　390

The new museum complex recreates battlements and earthworks and provides interpretation for the area, which consists of several sites spread over a valley that contains a small medieval town. And both of the two objects appear as non-obtrusive as possible in their respective contexts.

新建筑通过表皮材料来适应周边环境，造型上也与当地的建筑相仿，为人们再现了中世纪时期的城堡和小镇的一些意象。

A Canadian Museum in a Church Claire and Marc Bourgie Pavilion of Quebec and Canadian Art　400
蒙特利尔艺术博物馆：古老教堂中的艺术博物馆

Transition and continuation methods are meticulously used in designing this museum, the new buildings integrated into old buildings and its surroundings spontaneously.

该博物馆的设计以巧妙的过渡、延续手法，将新建筑自然地融入旧建筑中，并使得新馆与周围环境完美融合。

National Cowboy and Western Heritage Museum 美国西部牛仔历史博物馆 408

After this major expansion, the new entry to the museum is anchored by a sweeping curved canopy. This layout allows the facility an excellent forum for showing and integrating art with its architectural surroundings.

在改造的过程中，扩大了面积，并在博物馆入口处设计了弧形檐篷，艺术与建筑环境的更好的结合，使这里变成了一个极佳的交流场所。

Lightcatcher at the Whatcom Museum 沃特科姆博物馆莱特卡彻楼 416

The most visible and innovative feature—the lightcatcher—a multi-functional translucent wall that reflects and transmits the Northwest's most precious and ephemeral natural resource, sunlight which features an interpretive exhibit about the low-impact development strategies.

建筑耀眼的、创新式的透明墙面，能够捕捉美国西北部最宝贵的自然资源——阳光，同时也更好的诠释了生态设计理念。

Tel Aviv Museum of Art 特拉维夫艺术博物馆 422

A spiraling, top-lit atrium, whose form is defined by subtly twisting surfaces that curve and veer up and down through the building, serve as the surprising, continually unfolding vertical circulation system; while the natural light from above is refracted into the deepest recesses of the half-buried building.

建筑内有一个成螺旋状上升的明亮中庭，经过巧妙设计，形成惊人的垂直环流交通，并使得顶部光线直达建筑底层最深处。

Museum MUMAC Museum of Coffee Machine MUMAC 咖啡机博物馆 432

The facades of the museum are covered with strips of metal "red Cimbali", sinuous and enveloping to resemble the waves of hot coffee, which at night filters the artificial light creating a striking illuminated reticle that evokes the energy of MuMAC.

建筑外墙选用红色的带状装饰，象征热咖啡飘动的热气，加上背景光的照射，使其散发出引人注目的光芒。

Kamppi Chapel 康比教堂 442

The project widely uses wooden materials in the surface, interior walls and furniture, etc., to make a warm and quiet environment for the chapel.

项目大量采用木料，用于外立面、室内墙面、家具等，为教堂营造温暖安静的氛围。

Taiyuan Museum of Art 太原美术馆 452

Exterior light weight honeycomb panels with stone veneer produce an evocative and elusive material effect and the perception of an exceptional scale. The panels are reflective as if metallic.

外立面上的轻型蜂窝板贴有薄片石材镶面，如金属一般可以反光，产生出一种引人深思而又难以捉摸的材料效果。

Tampa Museum of Art Tampa, Florida 佛罗里达坦帕美术馆 460

A glass pedestal supports the jewelbox of art above. The building floats in the park, embracing it with its overhanging shelter and reflective walls. The building is not only in the landscape, but is the landscape.

建筑底部的玻璃基础使建筑整体像一个漂浮于公园之上的珠宝盒，不但很好地融入周边风景，并且创造出更为绚丽迷人的风景。

Urban Style
Dynamic Line
Open Space
Colorful Facade

都市气息

动感线条

通透空间

立面丰富

Shopping Mall
大型购物中心

With the rapid development of cities, large-scale shopping malls spring up and stand remarkably in hustle and bustle streets. As a kind of commercial complex integrating shopping, catering, recreation, entertainment, accommodation and travel together, more and more shopping malls become new landscapes of cities because of their excellent designs. To create a fashionable, dynamic and distinctive high-end commercial space is the goal for all designers.

Design for this kind of architecture usually pays much attention on details: facade style and material selection, interior style and decorating material selection, entrance design, openness and functions of the interior spaces, structure of the atrium or patio, etc. Excellent design will greatly increase values of the shopping mall. Designer should pay attention to every detail and then well show the economic, social and cultural value of the building.

随着城市的繁荣，大型购物中心像雨后春笋般在城市繁华的街道上拔地而起。大型购物中心是集购物、餐饮、休闲、娱乐、住宿和观光旅游为一体的商业综合体。越来越多的大型购物中心因其优秀的设计而成为城市的新风景。打造一个时尚、高端、动态但又有别于其他购物中心的商业空间成为设计师们的共同追求。

大型购物中心的设计往往需要在建筑外观的立面风格与材料选择、内部的装潢风格与装饰材料的挑选、购物中心的入口设计、室内商业空间的通透性和功能性的协调、中庭焦点广场的结构设计等方面下很大的功夫。好的设计方案往往是从多方面提升大型购物中心的存在价值，大到楼体造型，小到立面表层涂料都需要设计师精心设计，才能将大型购物中心的经济、社会、文化价值体现得恰到好处。

KEYWORD 关键词		
Skin and Facade 表皮与立面	Glass Facades 玻璃立面	
Materials 材料	Steels, Glass etc 钢材、玻璃 等	
	Greening Space 绿色空间	

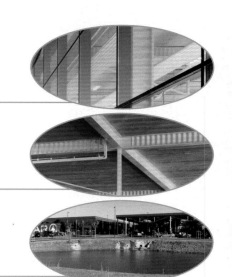

Location: Styria, Austria
Client: SES-Spar European Shopping Centers GmbH
Architect: Ernst Giselbrecht + Partner ZT GmbH
Project Team: DI Ingomar Platzer, DI Christian Liegl, DI Peter Fürnschuss
Plot Area: 19,800 m²
Net Area: 7,700 m²
Total Floor Area: 8,000 m²

项目地点：奥地利史蒂利亚
客户：SES-Spar European Shopping Centers GmbH
建筑设计：奥地利 Ernst Giselbrecht + Partner 设计事务所
项目团队：DI Ingomar Platzer, DI Christian Liegl, DI Peter Fürnschuss
占地面积：19 800 m²
净面积：7700 m²
总建筑面积：8000 m²

Interspar Fürstenfeld

Interspar Fürstenfeld 超市

Features 项目亮点

The architecture in no standard shape in general is a great solution to the connection between the retail area and the others, through a green and open design.

整体采用非标准对称的建筑形式，很好地解决了零售区与其他区域之间的联系，设计绿色、开放。

Overview 项目概况

Through its characteristic shape, Interspar Fürstenfeld demonstrates that it is no standard solution, instead drawing a reference to the structure of the plot of land. This is highlighted by the designed green space introduced between the projecting roof and retail area. The green space is a meeting point, serving as a filter between shop and parking lot. Together with terraces that invite one to dwell, it creates a high-quality ambiance on the threshold to shopping activities.

　　本案充分利用其独特的场地形状诠释了一项非标准的建筑形式，而不仅仅是作为一个参考。在向外伸出的屋顶和零售区之间设计了独具匠心的绿色空间，既是一个交汇点，也是商场和停车场的过渡区。绿色空间中设有休闲露台，在入口处营造了通往购物区的一个高质量氛围。

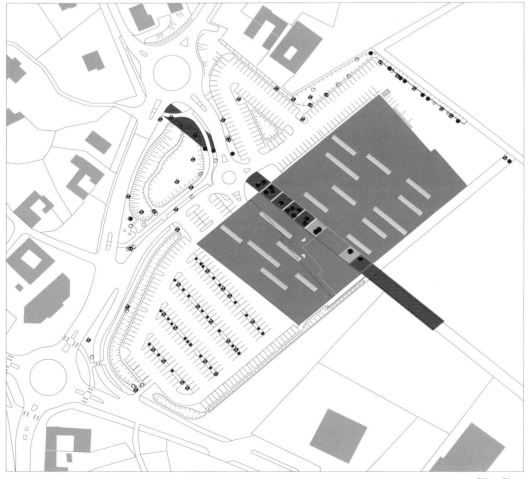

Site_Plan

Facades and Surface
立面与表皮

The supermarket hall is flooded with light, re-interpreting the concept of supermarket space through its design and open character. Particular focus was placed on the roof, the "5th façade", since it is visible from the historical town centre of Fürstenfeld. Thus the new market acts as a highly visibly indicator of Fürstenfeld's dynamic development as a town.

超市大厅灯火明亮,其别致开放的特点再次诠释了超市空间的设计理念。由于在历史悠久的Fürstenfeld城镇中心能够看到本项目的第五个立面——屋顶,因此设计师将设计重点放在屋顶,使其成为展示Fürstenfeld's城镇动态发展的引人注目的标志。

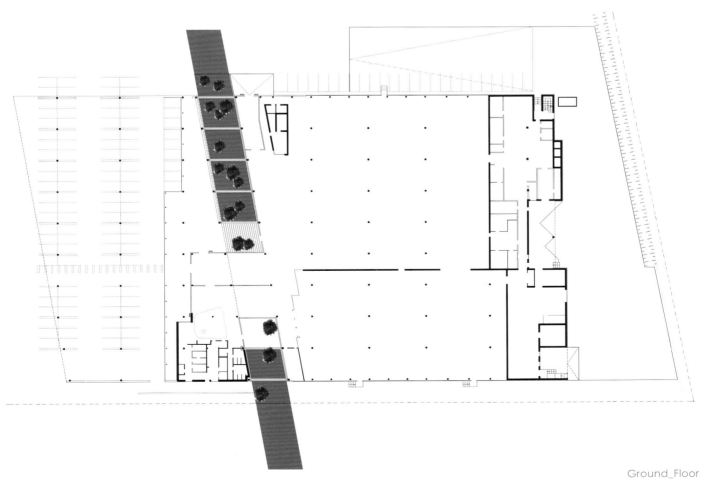

Ground_Floor

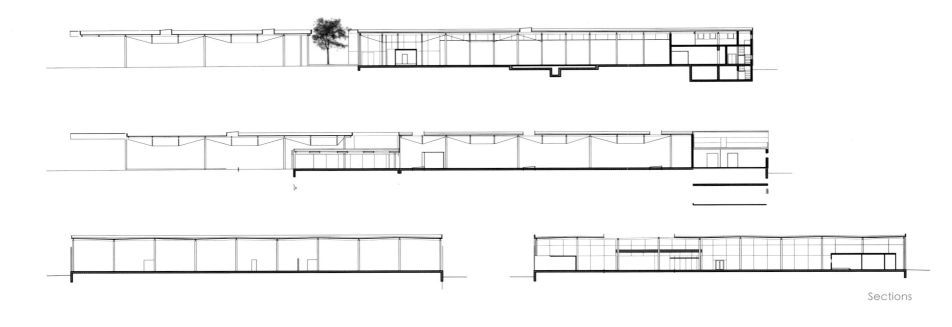

Sections

ANSICHT SUEDOST

ANSICHT NORDWEST

ANSICHT SUEDWEST

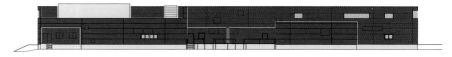

ANSICHT NORDOST

Elevations

Structure and Materials 结构与材料

The large transparent entrance wall makes it possible to display the wide product range. The cross-laminated boards have been used for the ceiling to resemble the traditional market, the generous ceiling height to achieve a spacious feeling and the intentionally provided green space for meeting and socialising - features which all contribute to a functional and pleasant atmosphere. The roof has been perforated by large window strips in order to maximise the use of natural day light, and as a result, the interior of the store is bathing in light. The giant canopy roof above the car park does not only shelter customers while loading groceries into their cars, it also provides sun-shading for the transparent, south-facing entrance façade.

透明的入口墙壁能够充分展示广泛的产品范围，天花板上的交叉层压板再现了传统的市场。天花板的高度营造了宽敞的感觉，特意设计的功能性绿色空间作为会面和社交活动场所，营造了一种愉悦氛围。为了最大限度的利用自然光线，屋顶开了多个大的条形窗口，商场内部也沐浴在自然光线下。停车场上方巨大的罩蓬屋顶不仅为搬着货物上车的顾客提供遮蔽，同时也为朝南的透明入口外墙遮挡阳光。

KEYWORD 关键词		
Skin and Facade 表皮与立面	Mixed facade 混合立面	
Materials 材料	Glass Panel, Steel, Etc. 玻璃面板、钢等	
Structure 结构	Reinforced Concrete 钢筋混凝土结构	

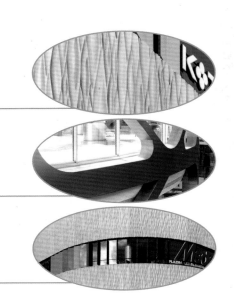

Location: Pendik-İstanbul/Türkiye
Owner: TSKBYO
Architecture Design: Erginoglu & Çalislar
Enclosed Area: 90,000 m²
Photographer: Cemal Emden

项目地点：土耳其伊斯坦布尔彭迪克
业主：TSKBYO
建筑设计：Erginoglu & Çalislar
总面积：90 000 m²
摄影师：杰马尔埃姆登

Pendorya Shopping Center
PENDORYA 购物中心

Features 项目亮点

The facade uses precast concrete panels and glass, while its interior design and form are derived from the structure's particular location.

立面采用预制混凝土板与玻璃的混搭模式，内部空间根据建筑结构的特定位置打造活力型购物空间。

Overview 项目概况

Pendorya Shoping Center is located on the E-5 highway, in a development zone, which is expected to have offices and residential buildings in the future. The main criteria used in determining the design and form are derived from the structure's particular location. Turning its back to the busy and noisy E-5 highway, dynamic and surprising inner spaces of the building are exposed to the sheltered open square, which is expected to become the new gathering area.

With direct access from car parking and inner spaces transferred to the outer spaces, this new open square is anticipated to become the new point of attraction for pedestrian access. It is envisaged as a twenty-four hour living area where different outdoor activities are pursued. Basement floors are reserved for car parking. Service and parking lot entrances are from the E-5 side for easy access. Flexibility in the inner space is achieved by variations in the plan schemes of the retail areas and easy pedestrian flow by a well-designed circulation system. In this way movement between the different common/social areas, such as the Open Square, covered square, terrace area, and galleries is facilitated.

PENDORYA 购物中心位于 E-5 高速公路旁的一个开发区内，将来这里还会开发办公楼和住宅等项目。设计方案中采用的主要标准是根据建筑结构的特定位置来制定的。大楼背面是嘈杂繁忙的 E-5 高速公路，外部是一个有遮蔽的露天广场。而商场内部则充满活力，处处带给人惊喜。PENDORYA 有望成为彭迪克新的中心商业区。

购物中心的停车场和商场内部有捷径直接通向外部空间，而这个独具特色的露天广场也将成为人行道上的新焦点。该中心实行 24 小时营业制，户外也会安排丰富的活动。内部空间的灵活性则通过销售区域多变的规划方案和精心设计的客流系统来实现。

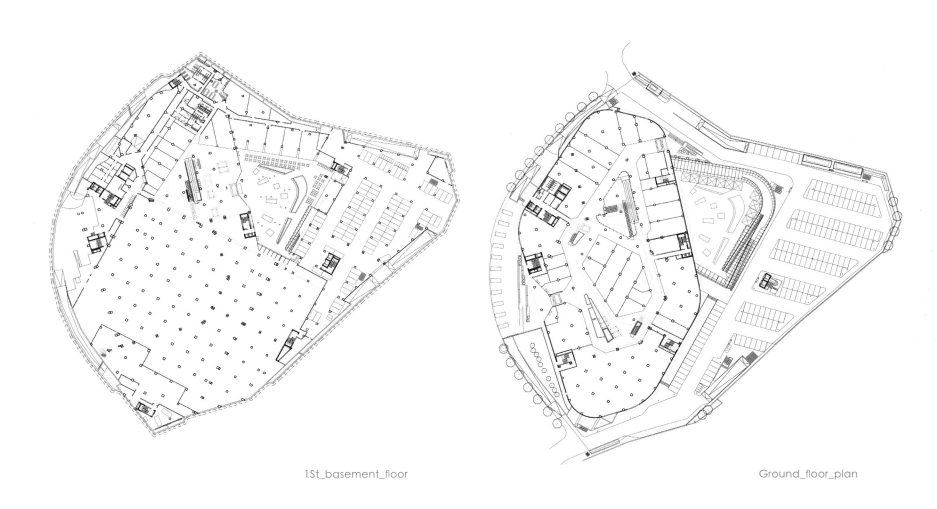

1St_basement_floor

Ground_floor_plan

Facade and Materials 立面与材料

Taking into consideration the intensity of the surrounding highway traffic, the façade is designed with a logo-strip to easily catch the eye. The façade is covered with glass panels, on the ground floor, whereas custom-made, precast concrete panels with patterns dress the upper floors of the façade. Long-term maintenance as well as visual coherence is thus achieved.

建筑外墙覆盖有玻璃面板，地面层的设计为顾客量身打造，印有图案的预制混凝土板用来装饰外墙的上半部分。同时，商场还坚持长期的维护以确保维持该购物中心的视觉效果。

DEVELOPMENT SCHEMES

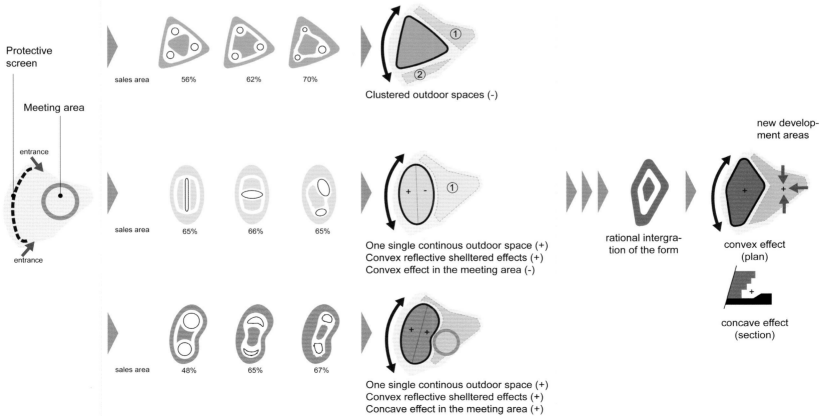

SITE PLAN
Visual perception

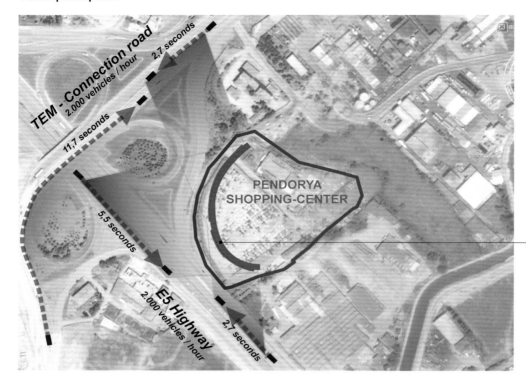

Protective screen

SITE PLAN
Relations with neighbourhood

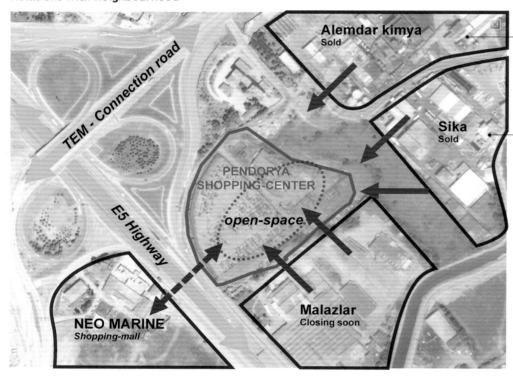

Possible urban development and transformation areas

Possible urban development including pedestrian zone

SITE PLAN
Environment

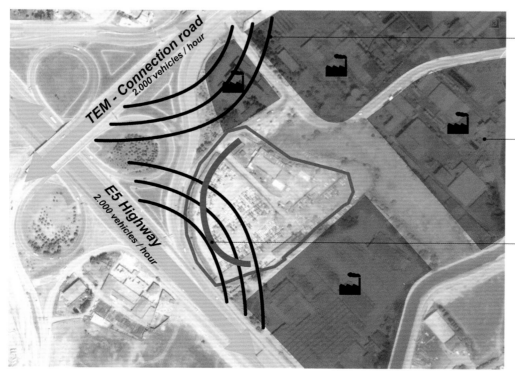

High perception
Disturbance due to noisy traffic

No pedestrian connections between areas

Protective screen

INTERIOR SPACE
Square

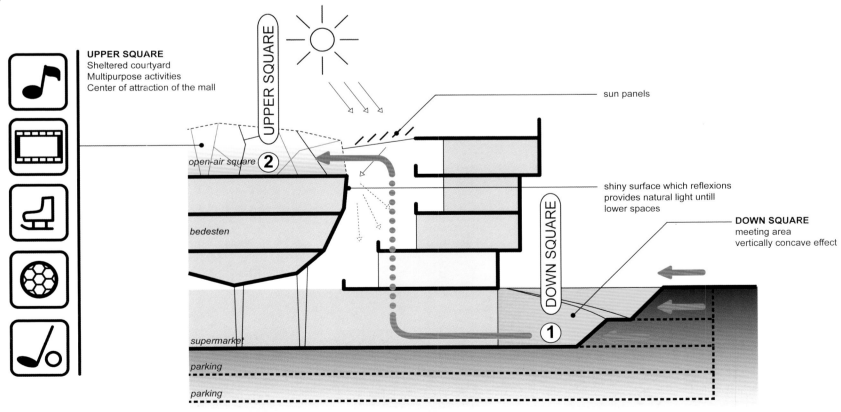

UPPER SQUARE
Sheltered courtyard
Multipurpose activities
Center of attraction of the mall

sun panels

open-air square

shiny surface which reflexions provides natural light untill lower spaces

DOWN SQUARE
meeting area
vertically concave effect

bedesten

supermarket

parking

parking

SQUARES / OPEN SPACES

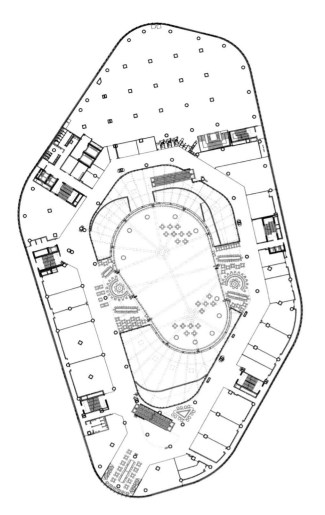
2Nd_floor_plan

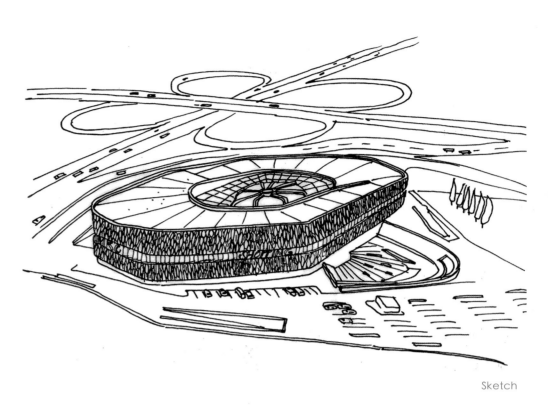
Sketch

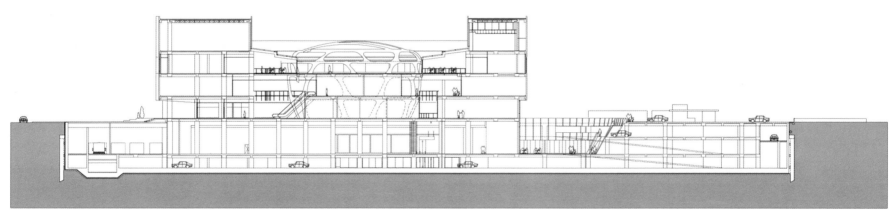
Section1

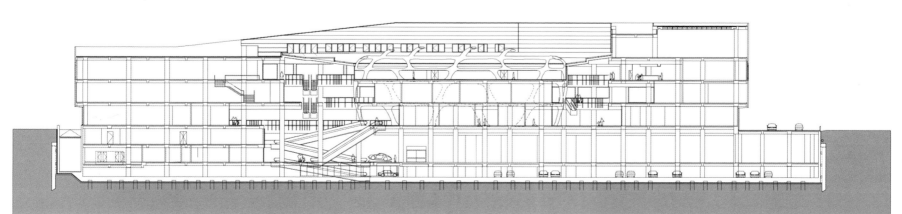
Section2

KEYWORD 关键词	Skin and Facade 表皮与立面	Glass Facade 玻璃立面
	Materials 材料	Glass, Steel, Etc. 玻璃、钢等
	Shape 造型	Dynamic Facades 动态开放造型

Location: Moscow, Russia
Client: Capital Partners
Design Firm: RTKL Associates Inc.
VP-in-Charge: Alan Morgan
Size: 3,444,448 SF / 320,000 SM
Photographer: David Whitcomb

项目地点：俄罗斯莫斯科
客户：资本合伙人
建筑设计：RTKL
项目负责人：艾伦·摩根
建筑面积：320 000 m²
摄影师：大卫·惠特科姆

Metropolis
大都会

Features 项目亮点

The design of Metropolis mimics an open-air street scene through various architecture details to create a pop, sophisticated and dynamic shopping space.

设计模拟出一个露天的街道场景，通过一系列的建筑结构细部的表现，营造出一个时尚、高端、动态的商业空间。

Overview 项目概况

The design team for Metropolis had two goals: to introduce a modern mixed-use environment that supports around-the-clock activity while also creating a sophisticated space that fits into the local setting.

项目设计要达到两个目标：既要体现出现代的商业综合体的不夜城性质，又要创造了一个符合当地风格的高档场所。

Shape and Structure 造型与结构

Inspired by the energy of the urban environment, the design of Metropolis mimics an open-air street scene. Retail corridors that resemble shopping boulevards employ colors, patterns and finishes to transform individual storefronts into dynamic facades.

Various architecture details continue the aesthetic of an outdoor lifestyle center including a grand plaza and garden that form a junction between the retail component and three office buildings. With close proximity to public transportation, Metropolis is a celebrated extension of downtown Moscow.

　　大都会的设计灵感来自于城市活力，模拟设计出一个露天的街道场景。购物街的零售走廊利用颜色、图案和装饰将各个店面打造成动态的外型。

　　多个设计细节，包括在零售区和三幢写字楼交汇处的大广场和花园来丰富户外生活体验。大都会公共交通便利，可直达莫斯科市城区。

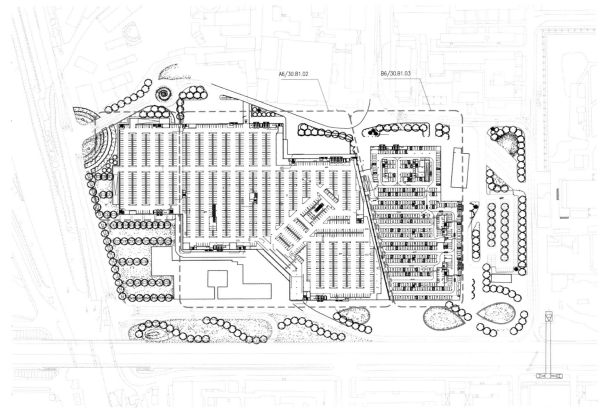
Basement

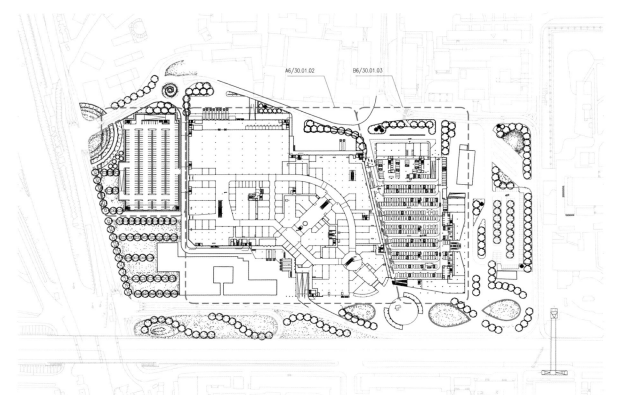
Level 1

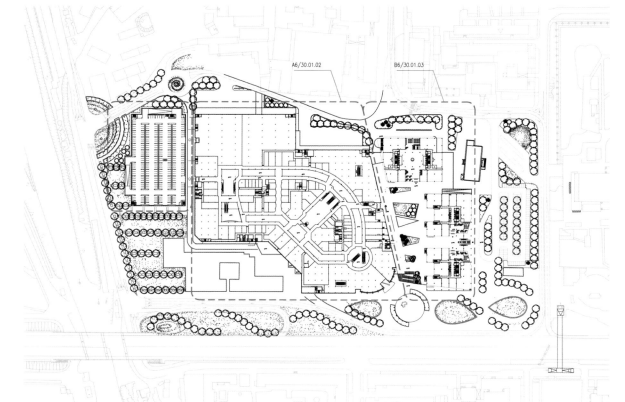
Level 2

Retail Section

Section

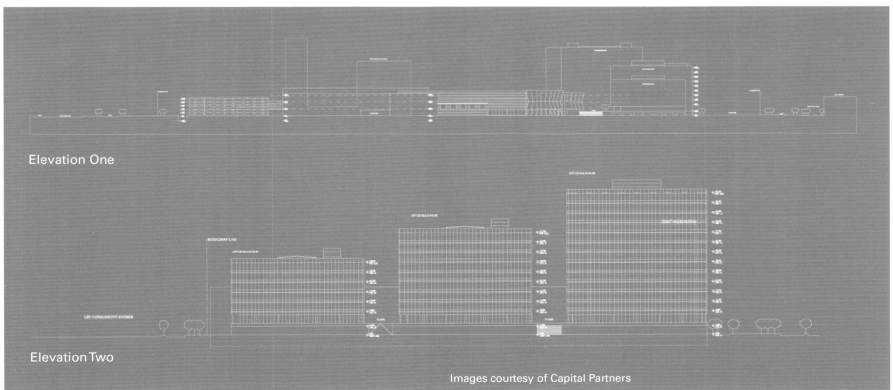

Elevation One

Elevation Two

Images courtesy of Capital Partners

Elevation

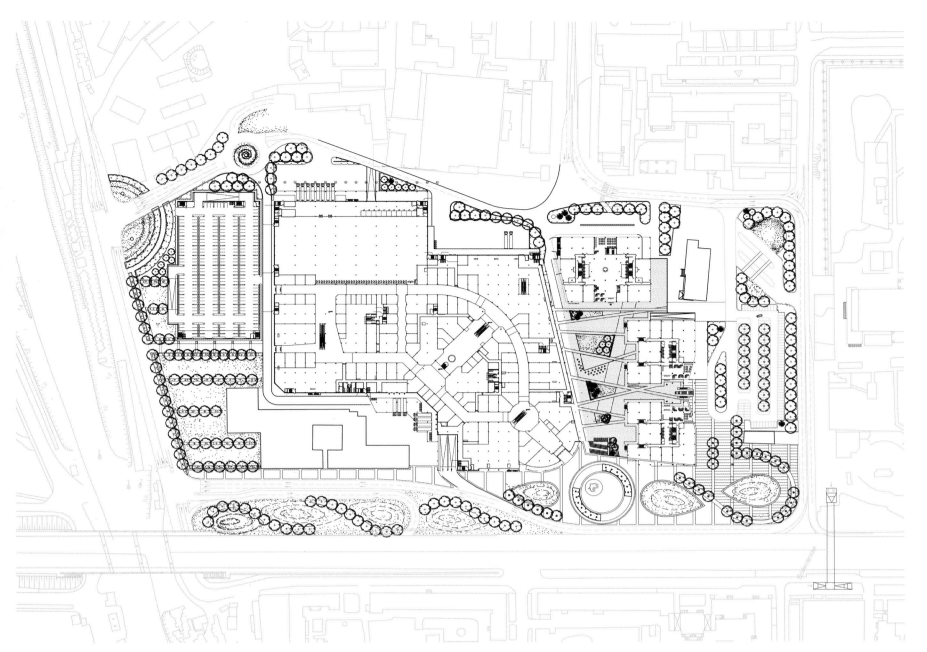

Metropolis_siteplan

KEYWORD 关键词	Skin and Facade 表皮与立面	Mixed Facade 混合立面
	Materials 材料	Glass, Metal, Etc. 玻璃、金属等
	Structure 结构	Prefabricated Concrete 预制混凝土结构

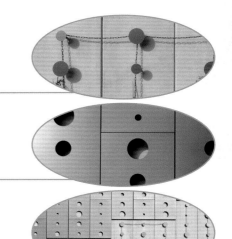

Location: Ljubljana, Slovenia
Client: Gradis G. Group, Mercator d.d.
Architectural Design: OFIS
Area: 3,400 m²
Photo: Tomaz Gregoric

项目地点：斯洛文尼亚卢布尔雅那
客　　户：Gradis G. Group, Mercator d.d.
建筑设计：OFIS
总建筑面积：3 400 m²
图片来源：Tomaz Gregoric

Dot Envelope
Dot Envelope 商场

Features 项目亮点

The skin uses metal and concrete with a series of irregular symmetrical dot decoration to create a great visual aesthetic appearance.

建筑表皮采用金属与混凝土相结合的构造，通过一系列不规则对称的圆点装饰设计，打造了一个极具视觉美感的商场外观。

Overview 项目概况

The existing site is listed as a historic industrial area containing the buildings of an old butchery complex, including a water-tower and an old butcher hall.

该项目的前身是一个旧的屠宰场，已被列为历史工业基地，包括一个水塔和一个旧的屠宰大厅。

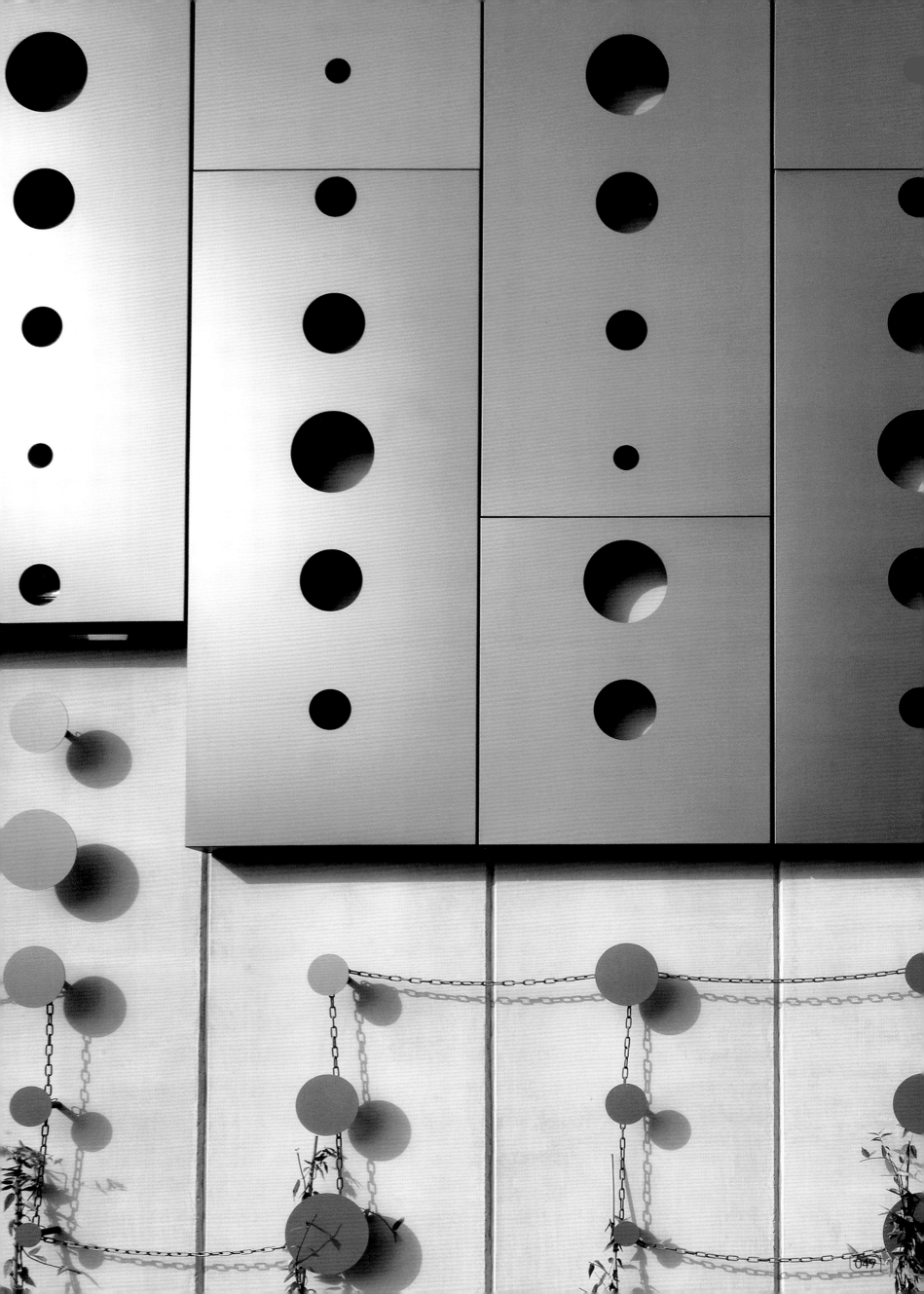

Siteplan

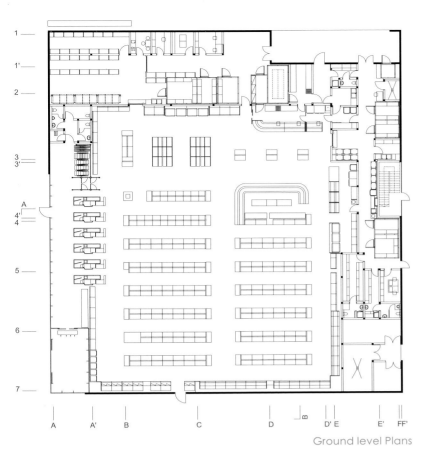
Ground level Plans

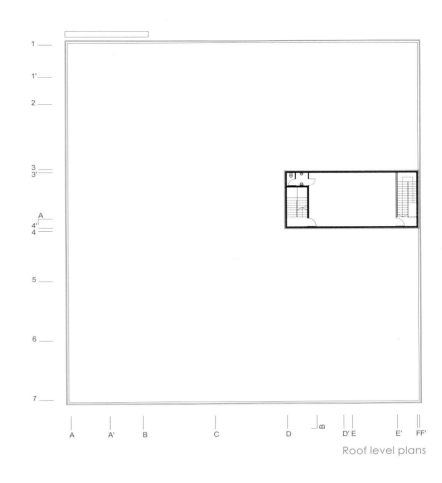
Roof level plans

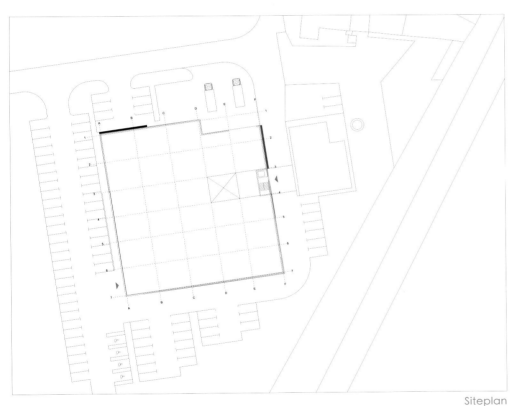
Siteplan

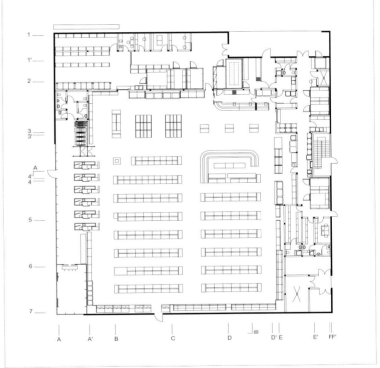
Plan A

Skin and Structure 表皮与结构

With parking place around its three sides, the centre is easy to access. What worth to tell is that less than 1/3 of the renovation budget is been used to decor these three parking lots.

The interior and all internal finishing were designed according to the "Mercator standard" – as prescribed by the shopping company's chain for all their standardized malls. In time, the surface will be covered with green climbing plants and it will become eco-green.

新购物中心的3面都设有停车场，简单实用。其中，尤为重要的是3面停车场的装修费用却没有超过1面停车场的装修预算。所有内部装修和装饰都按照"墨卡托标准"严格执行，即购物公司对所有商场制定的统一标准。

随着时间的推移，外墙绿色的攀爬植物还会覆盖建筑表面，起到绿化效果。

EAST

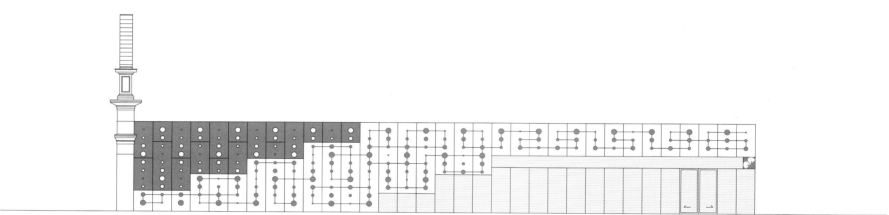
WEST

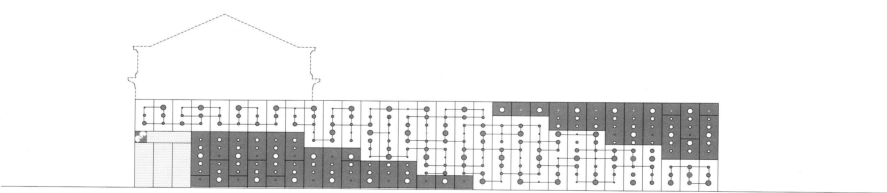
SOUTH

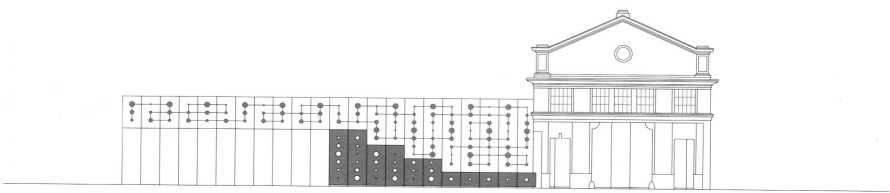
NORTH

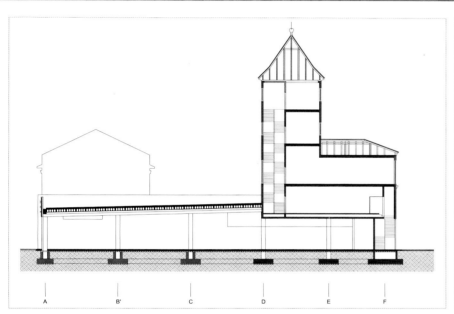

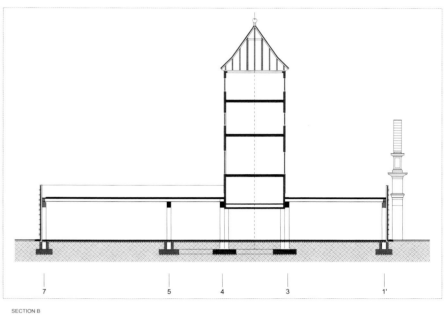

SECTION A

SECTION B

KEYWORD 关键词	Skin and Facade 表皮与立面	Glass Facade 玻璃立面
	Materials 材料	Glass, Stone, Metal, Etc. 玻璃、石材、金属等
	Shape 造型	Canyon-Style Curve 峡谷流线造型

Location: Seoul, South Korea
Client: GS Engineering and Construction
Project Design: The Jerde Partnership
Site Area: 26,400 m²
Floor Area: 294,600 m²
项目地点：韩国首尔
客　　户：GS 工程建设集团
建筑设计：美国捷得国际建筑师事务所
占地面积：26 400 m²
建筑面积：294 600 m²

Mecenatpolis

Features 项目亮点

Connecting to the residential and office buildings, the canyon-style mall also provides a dynamically integrated pedestrian place.

峡谷造型的商业空间作为纽带，既连接着住宅、办公空间，也营造了一个流动式的步行体验。

Overview 项目概况

Located in the evolving Hapjeong neighborhood in downtown Seoul, adjacent to the Hapjeong train station, the 295,615 square-meter Mecenatpolis is a transit-oriented destination featuring a composition of three high-rise luxury residential towers, one high-rise class-A office tower, and an expansive, open-air mixed-use public realm at its base, inspired by the elegant form and curve of a natural canyon.

项目坐落在首尔市合井商业中心地段，毗邻合井火车站，Mecenatpolis 建筑面积为 295 615 m²，包括三个高层住宅楼，一个高层 A 级办公大楼和带有影院、多功能表演厅、折扣店的开放式商业综合体。设计灵感来自于天然大峡谷的高雅和曲线。

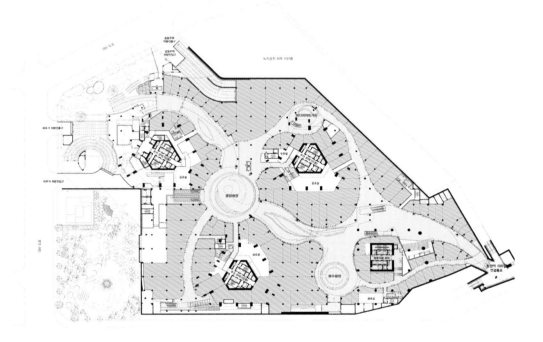

Structure and Materials
结构与材料

The open-air village comprises nearly 90,000 square meters of retail and dining, as well as an events auditorium and public park for recreational and cultural activities where visitors and residents can gather and engage. Terraced balconies, glass bridges, landscaped roof gardens, open-air spaces, mesmerizing water features and a grand central plaza all reinforce the idea of a fluid and choreographed circulation system at the core of the retail complex. The free-flowing rhythm incorporates the region's age-old concept of a 'unified well' to shape the main central plaza that is Mecenatpolis' core gathering area, which showcases an active water feature and an abundance of open-air public space and restaurant seating areas, designed to draw people inward. A multi-purpose hall, designed as an iconic lantern piece with beautiful wood paneling, sits adjacent to the cultural park and will host events such as concerts and other special performances.

The organic design of the complex evolved by reinforcing key connections through the site from the station to the surrounding neighborhoods, creating a fluid pedestrian experience within the retail, while making the best use of the location in this emerging neighborhood next to one of Seoul's busiest transportation hubs. Realizing the highly active nature of the city's residents, the new urban destination celebrates the outdoor experience as a retreat from the fast pace of the surrounding city—an urban park that will stimulate the resurging community while providing a dynamically integrated pedestrian place, authentic to the culture of Seoul.

　　露天广场包括近90 000 m² 的餐饮零售以及文化娱乐空间。人们聚集于这里的礼堂、公共花园、梯田式阳台、玻璃桥、绿化屋顶的花园及露天场所，感受迷人的水景和人来人往的新城市中央广场零售区。中心广场是Mecenatpolis的核心聚集区，自由流动的节奏结合该地区古老的概念"统一"，这里的亲水区设有宽敞的露天公共空间和餐厅，设计将人们引向室内。有着一个标志性灯笼和一块美丽的木镶板的多用途礼堂，毗邻文化公园做举办各种活动使用，如音乐会和其他特殊活动。

　　利用先进技术，在Mecenatpolis附近的位置连接到首尔最繁忙的交通数据中心，使人流交通状况得到智能控制，在零售综合区创建一个流动式的步行体验，最好的实现居民高质量的城市生活。商业区的各种设施和服务让人们感受到正宗的首尔文化。

2F 영화관과 다목적 공연장으로 조성된 라운지몰
멀티플렉스 영화관과 다양한 공연 및 행사를 진행 할 수 있는 공연장 및 다양한 매장으로 구성되어 있습니다.

1F 트렌디한 로드샵들이 이어지는 가로 대면몰
쇼윈도우가 넓어 노출성이 뛰어난 매장이 길을 따라 스트리트몰로 구성되어 있습니다.

B1 테마거리를 따라 이어지는 감각적인 스트리트몰
수변공간과 조경시설이 어우러진 테마거리로 고객의 취향에 따라 선택이 편리한 매장으로 구성되어 있습니다.

B2 대형할인마트로 구성된 패밀리 라이프마켓
약 33,500㎡의 홈플러스가 자리 잡고 있어 높은 집객효과를 거둘 수 있습니다.

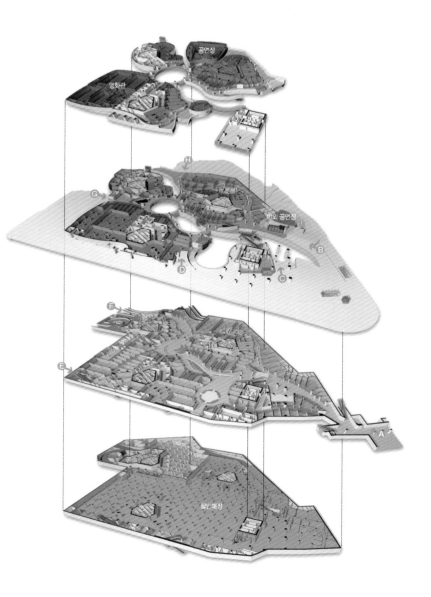

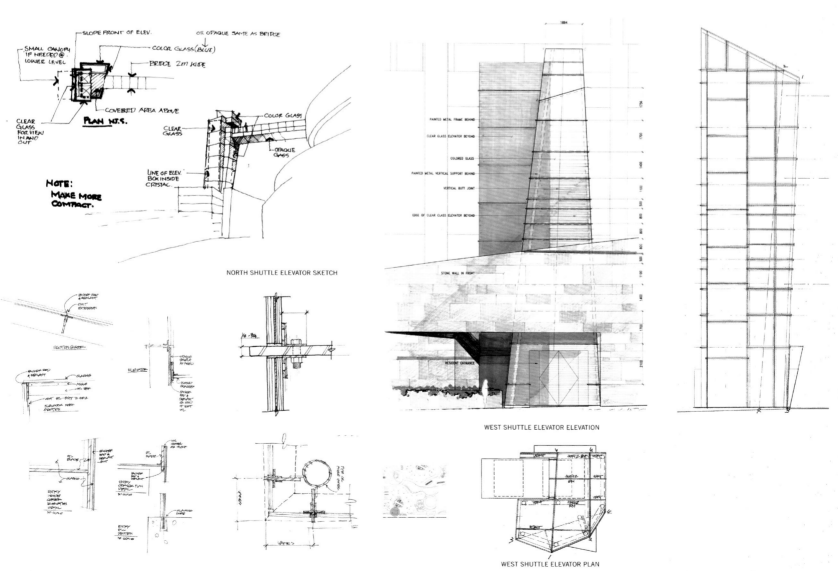

NORTH SHUTTLE ELEVATOR SKETCH

WEST SHUTTLE ELEVATOR ELEVATION

WEST SHUTTLE ELEVATOR PLAN

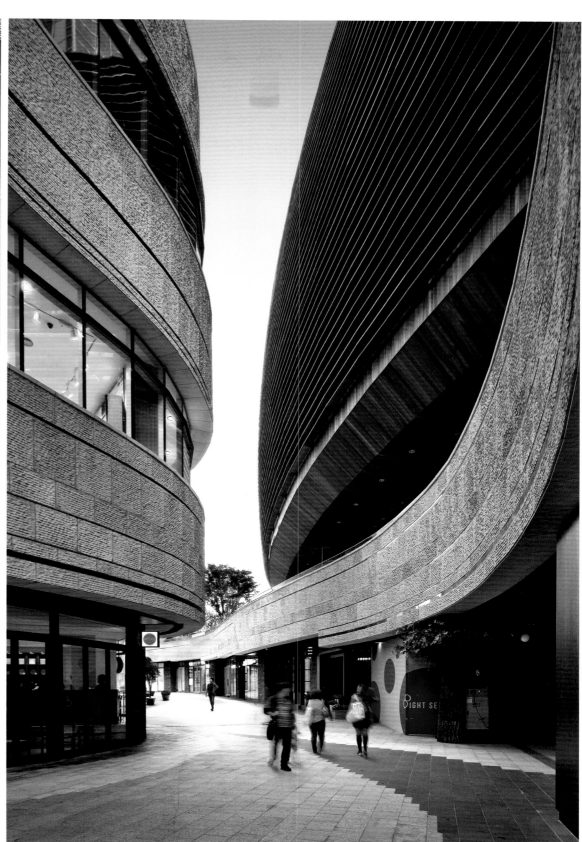

KEYWORD 关键词	Skin and Facade 表皮与立面	Mixed Facade 混合立面
	Materials 材料	Glass, Steel 玻璃、钢
	Shape 造型	A Folded Piece of Paper 折叠纸张造型

Location: Schladming, Austria
Architects: HOFRICHTER-RITTER Architects
Effective Area : 18,500 m²
Photographs: Paul Ott

项目地点： 奥地利施拉德明
建筑设计： 奥地利 HOFRICHTER-RITTER 事务所
有效面积： 18 500 m²
摄影： Paul Ott

Skiterminal – Schladming
施拉德明滑雪度假综合体

Features 项目亮点

The complex is an oversized loop composed of three independent but connected buildings, with a unique mixed facade of glass and metal materials.

整个综合体外部表现为一个环形结构，三个主要的建筑群既相互独立又互为整体，玻璃与金属材质的混搭表皮显得新颖而独特。

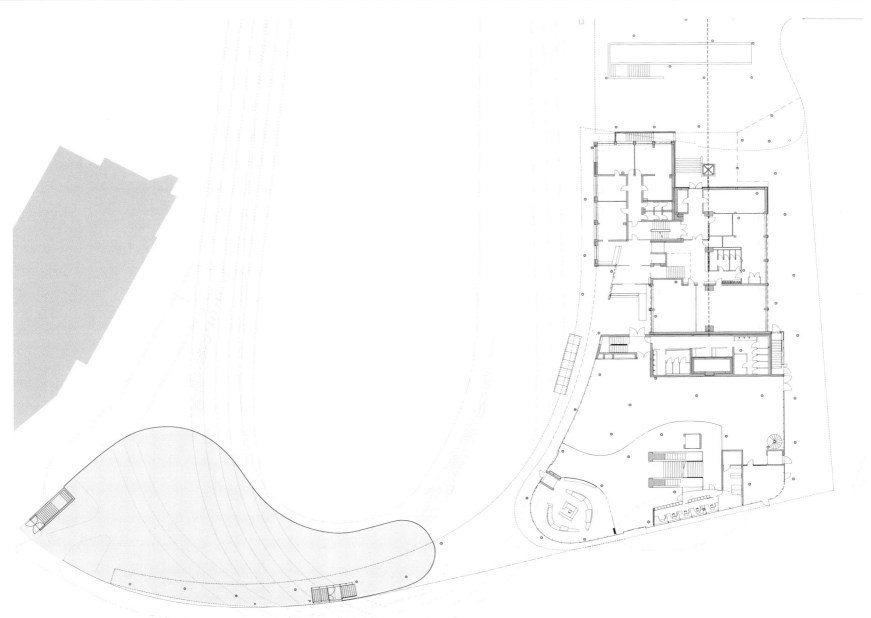

Groundfloor

DRAUFSICHT

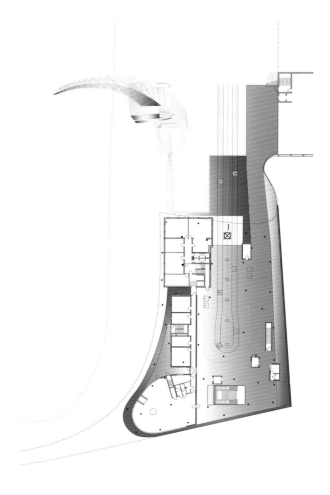

SEILBAHNHALLE

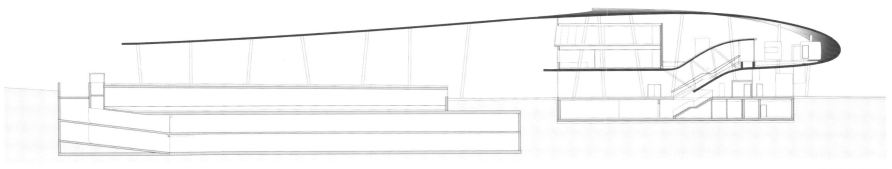

SCHNITT

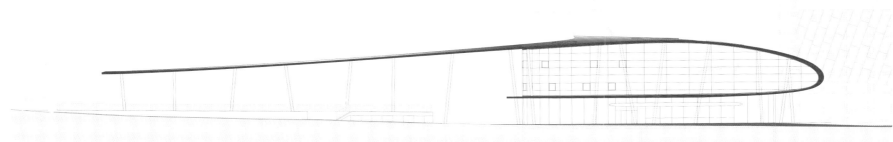

ANSICHT NORD

Shape and Structure　造型与结构

An oversized "Loop" (based on the model of a folded piece of paper) coats the whole building and is carried by 74 diagonal columns. Each and every single column can carry 230 tons. Inside the "Loop", three connected complexes of buildings are situated. In the Southern most part, the new headquarters of the Planai can be found. In the central building, offices for important partners such as Wintersportverein Schladming, Austrian Ski Federation or FIS are available. The Northern complex with 1,000 m² of glass facade is accessible for guests.

The concept of the "One-Stop-Shop" is new. At this central place, the holiday guests can find EVERYTHING they need for a comfortable and relaxing vacation.

由 HOFRICHTER-RITTER 事务所设计的滑雪商业综合体项目呈现出一个巨大的环形结构（犹如一个折叠的纸张模型）环绕整个建筑，同时由 74 根倾斜的柱子支撑，每根柱子能承重 230 吨。这个环形结构内部包含三个相连的建筑群，最南侧为 Planai 的新总部，中间建筑是重要合作伙伴如施拉德明冬季体育协会、奥地利滑雪联合会和 FIS 等的办公空间，北部的建筑拥有 1 000 m² 的玻璃外墙，对游客开放。

项目创新性地提出了"一站式度假商店"的概念，在中心场地上，度假的游客可以找到他们需要的任何东西，享受他们舒适轻松的假期。

KEYWORD 关键词	Skin and Facade 表皮与立面	Glass Facade 玻璃立面
	Materials 材料	Polycarbonate and Glass 聚碳酸酯与玻璃
	Shape 造型	Triangular Unit 呈三角形单元模块

Location: Gyonggi-Do, Korea
Architectural Design: JOHO Architecture
Designer: JeongHoon Lee
Site Area: 853.7 m²
Building Area: 639.4 m²
Floor Area: 2,554.29 m²
Photography: Sun Namgoong, Jeonghoon Lee

项目地点：韩国京畿道
建筑设计：JOHO 建筑设计事务所
设计师：JeongHoon Lee
场地面积：853.7 m²
建筑占地面积：639.4 m²
建筑面积：2 554.29 m²
摄影：Sun Namgoong, Jeonghoon Lee

Herma Parking Building
赫尔玛停车大楼

Features 项目亮点

The Herma parking building maximized the potential commercial uses of the land, maximized of lawful commercial area, the possibility of using front terrace, Increased the architectural value through front-skin design.

赫尔玛停车大楼的设计最大化地发挥了商业用地的潜力，最大限度的优化了商业空间，并通过建筑表皮的设计提升了建筑价值。

Overview 项目概况

The Herma parking building was started from a fundamental consideration of current parking lots in Korea. Maximize the potential commercial uses of the land. Maximize of lawful commercial area, the possibility of using front terrace. Compose of the parking ramp and the traffic line. Increase the architectural value through front-skin design.

赫尔玛停车大楼的设计考虑了当前韩国停车大楼的设计，其设计最大化地发挥了商业用地的潜力，最大限度的优化了商业空间并尽可能地使用阳台。考虑到停车场及交通线的组合，通过建筑表皮设计提升建筑价值。

Construction Process

CIP Piles + Mat Foundation

Column

Ramp

Parapet

Core

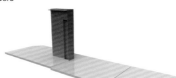

Girder

Slab

Wall + Glass

Rooftop Plant

Galvanized Pipe Frame

Polycarbonate Panel

Stainless Louver Frame

Steel Stairs

Aluminium Perforated Panel Frame

Galvanized Pipe + Paint

Stainless Pattern Frame

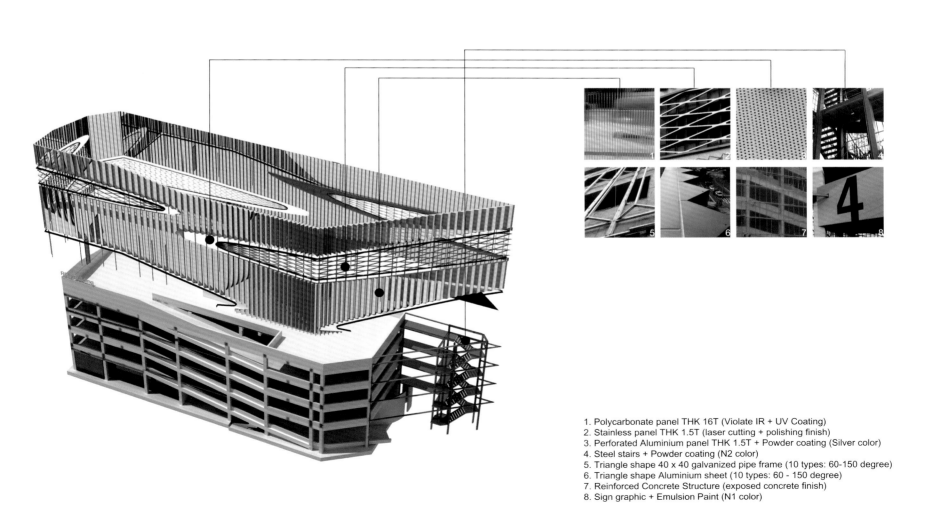

1. Polycarbonate panel THK 16T (Violate IR + UV Coating)
2. Stainless panel THK 1.5T (laser cutting + polishing finish)
3. Perforated Aluminium panel THK 1.5T + Powder coating (Silver color)
4. Steel stairs + Powder coating (N2 color)
5. Triangle shape 40 x 40 galvanized pipe frame (10 types: 60-150 degree)
6. Triangle shape Aluminium sheet (10 types: 60 - 150 degree)
7. Reinforced Concrete Structure (exposed concrete finish)
8. Sign graphic + Emulsion Paint (N1 color)

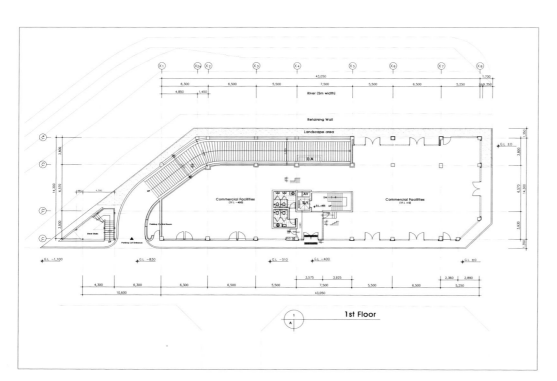

1st Floor

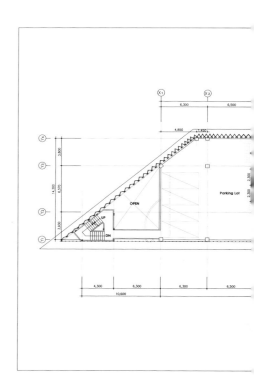

Skin and Materials 表皮与材料

The front part facing the river is a triangular unit. Contrarily the back surface of commercial area was finished with a flat surface. The polycarbonate used in this project has 5 layers. The external part was coated with violet colour, and the internal part with white colour. Additionally the external part was made with IR Coating and Coating surface treatment, which exerted such a feeling as the reflective surface of glass or metal depending on the angle of light. In the morning with the sun rising, it receives direct ray whose colon of outer cover minutely appears to be white, and at noon indirect one whose colour purely violet. In the afternoon with the sun setting, it reflects the evening glow and turns to be golden, while in the evening it reflects the internal lighting and external neon sign and directs unusual scenes.

建筑的正面呈三角形单元模块。背面商业空间的表面为平整表面。该项目其中五层采用了聚碳酸酯。建筑外观呈紫罗兰色，室内为白色。外观采用红外镀膜以及表面镀膜处理技术，该技术的采用强调了玻璃表面的反光效果。时间不同外观呈现的颜色也不同，早上太阳刚升起来时呈白色，中午呈纯紫色，下午呈现金黄色，到了晚上，在室内灯光以及室外霓虹灯的反射作用下建筑呈现美轮美奂的场景。

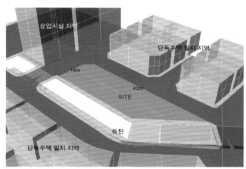

- Maximizing the potential commercial uses of the land

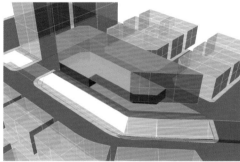

- Maximization of lawful commercial area, the possibility of using front terrace

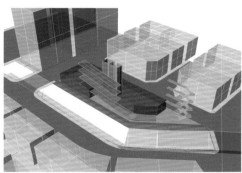

- Composition of the parking ramp and the traffic line

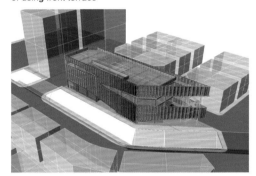

- Increasing the architectural value through front-skin design

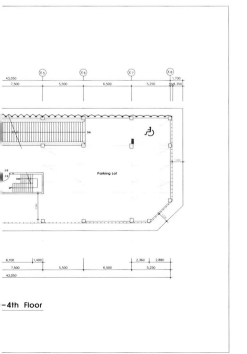

-4th Floor

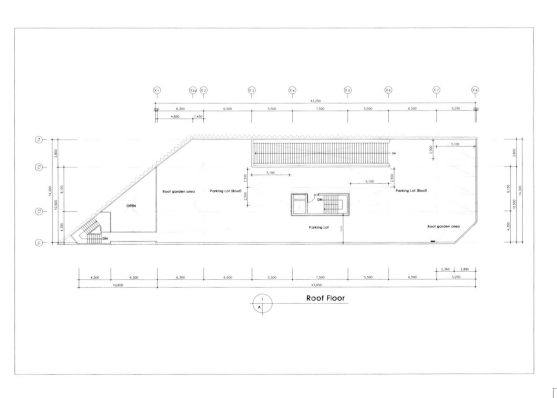

Roof Floor

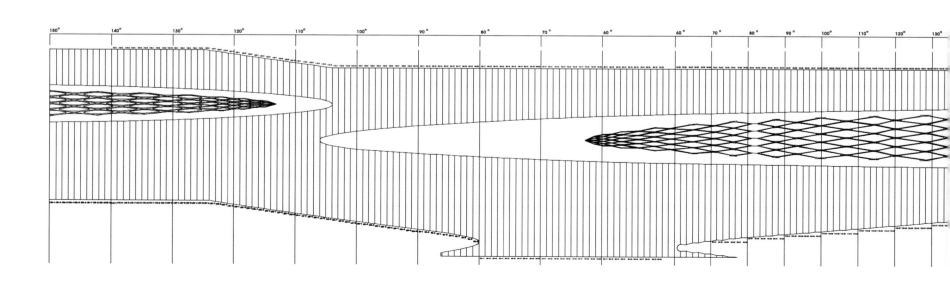

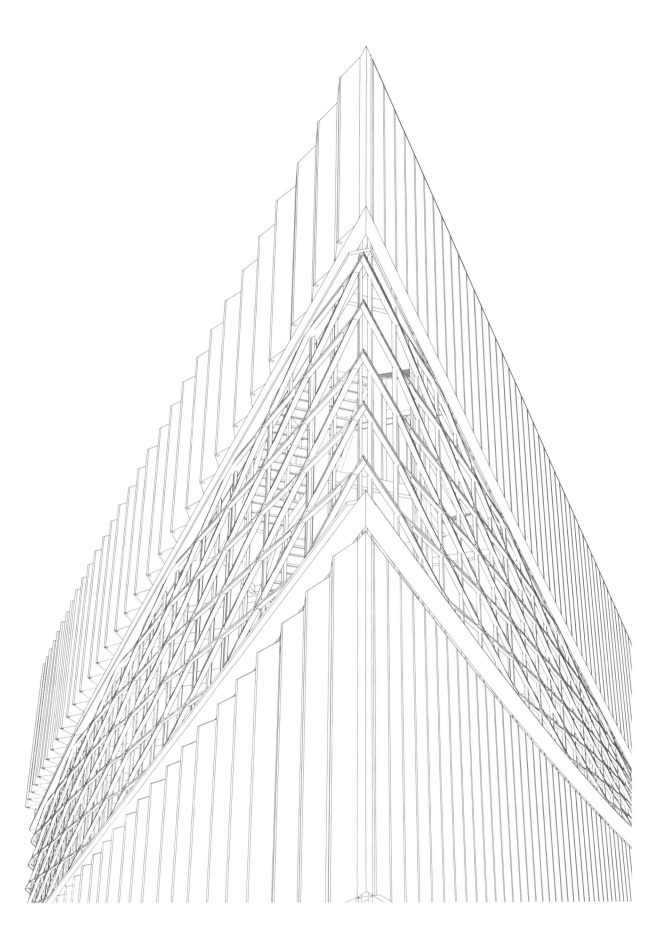
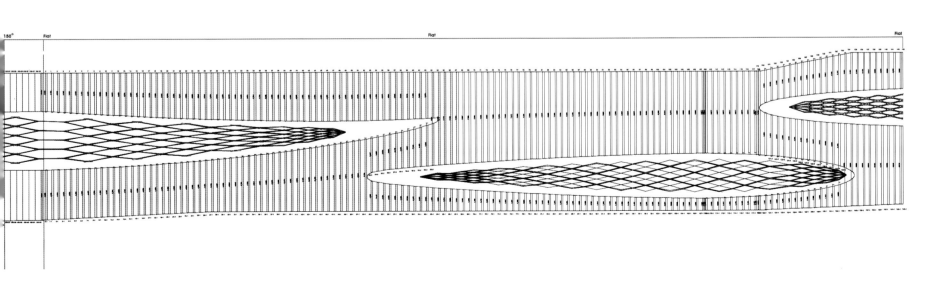

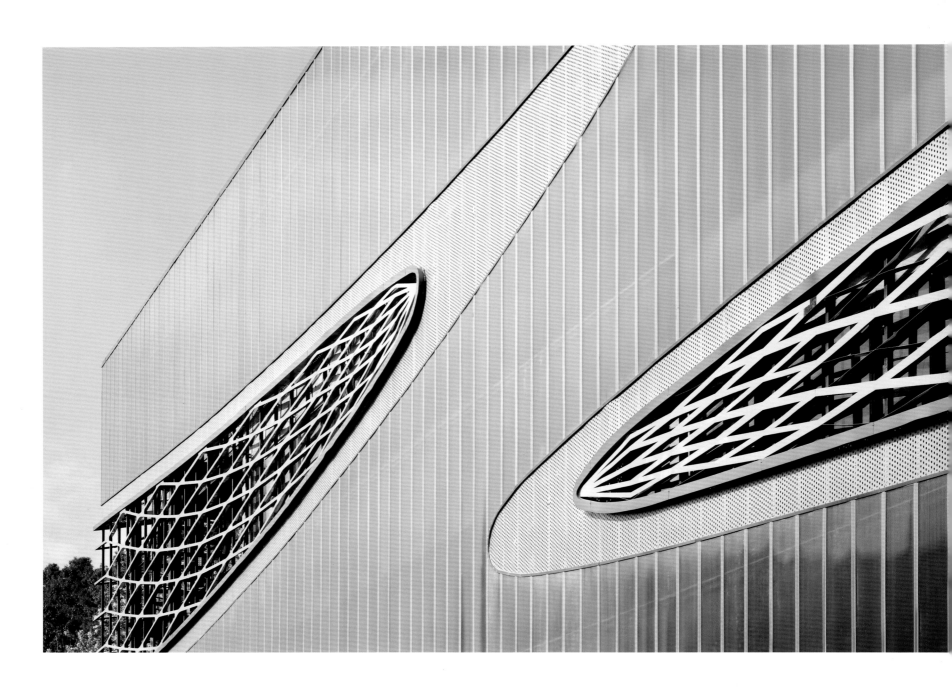

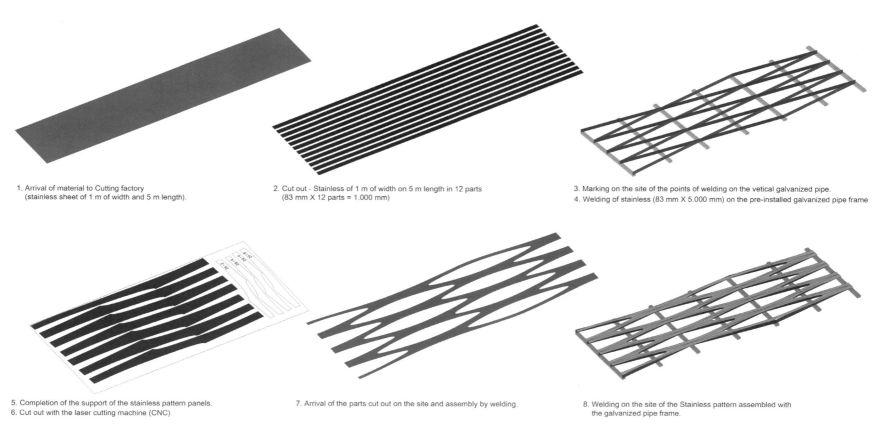

1. Arrival of material to Cutting factory (stainless sheet of 1 m of width and 5 m length).
2. Cut out - Stainless of 1 m of width on 5 m length in 12 parts (83 mm X 12 parts = 1.000 mm)
3. Marking on the site of the points of welding on the vetical galvanized pipe.
4. Welding of stainless (83 mm X 5.000 mm) on the pre-installed galvanized pipe frame
5. Completion of the support of the stainless pattern panels.
6. Cut out with the laser cutting machine (CNC)
7. Arrival of the parts cut out on the site and assembly by welding.
8. Welding on the site of the Stainless pattern assembled with the galvanized pipe frame.

fabrication process

pattern bottom with dim

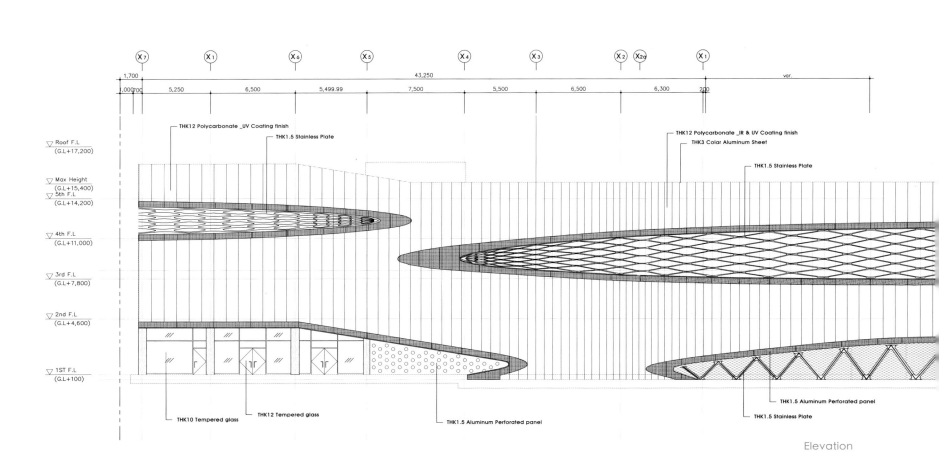

Elevation

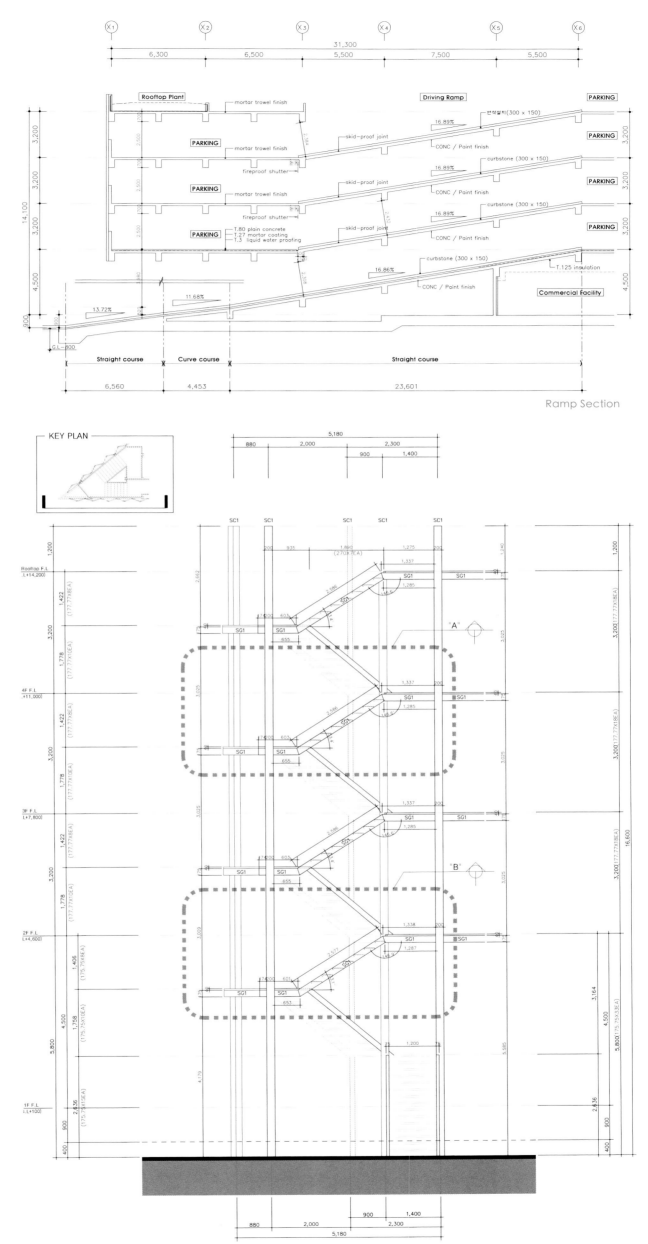

Ramp Section

Steel Stair Section

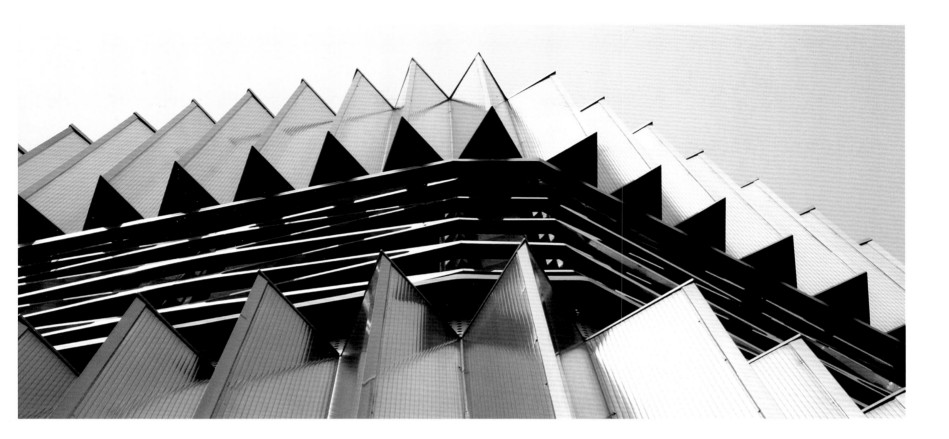

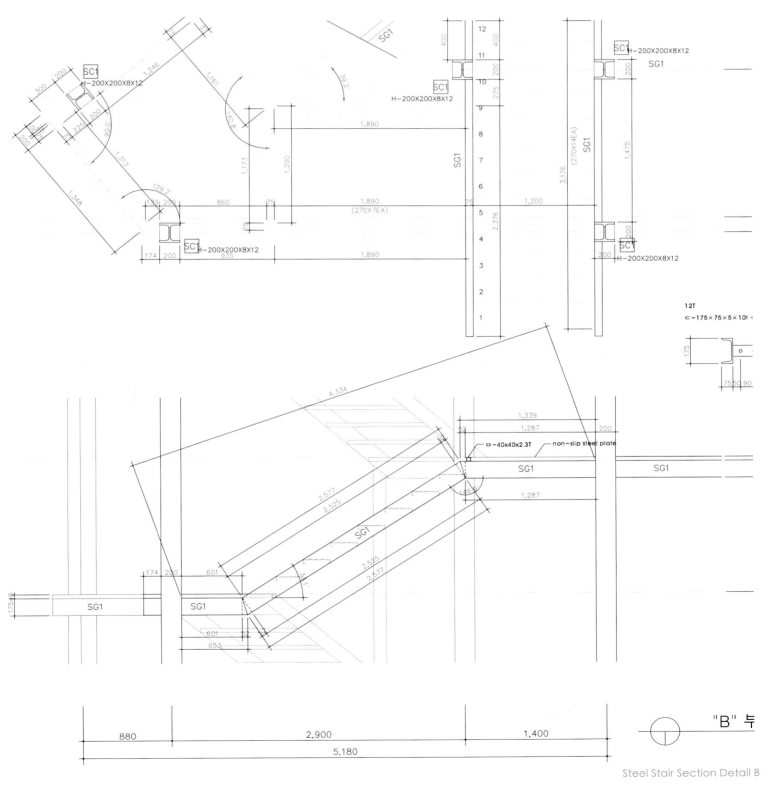

Steel Stair Section Detail B

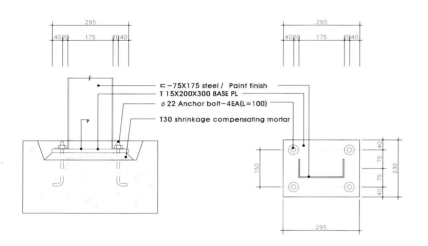

A | Column base detail

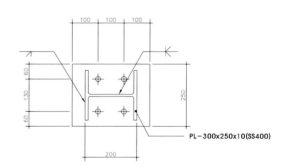

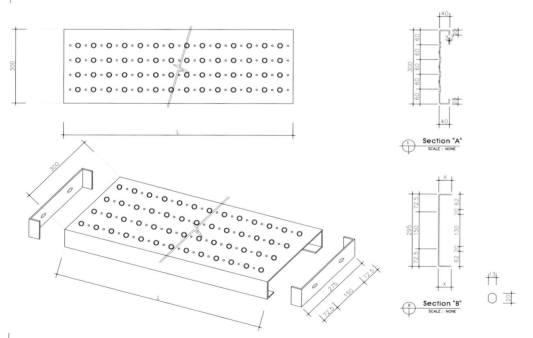

B | Floordeck detail

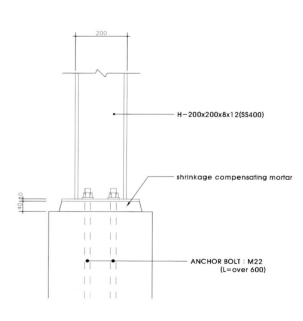

C | SC1 Column base detail

Base Detail

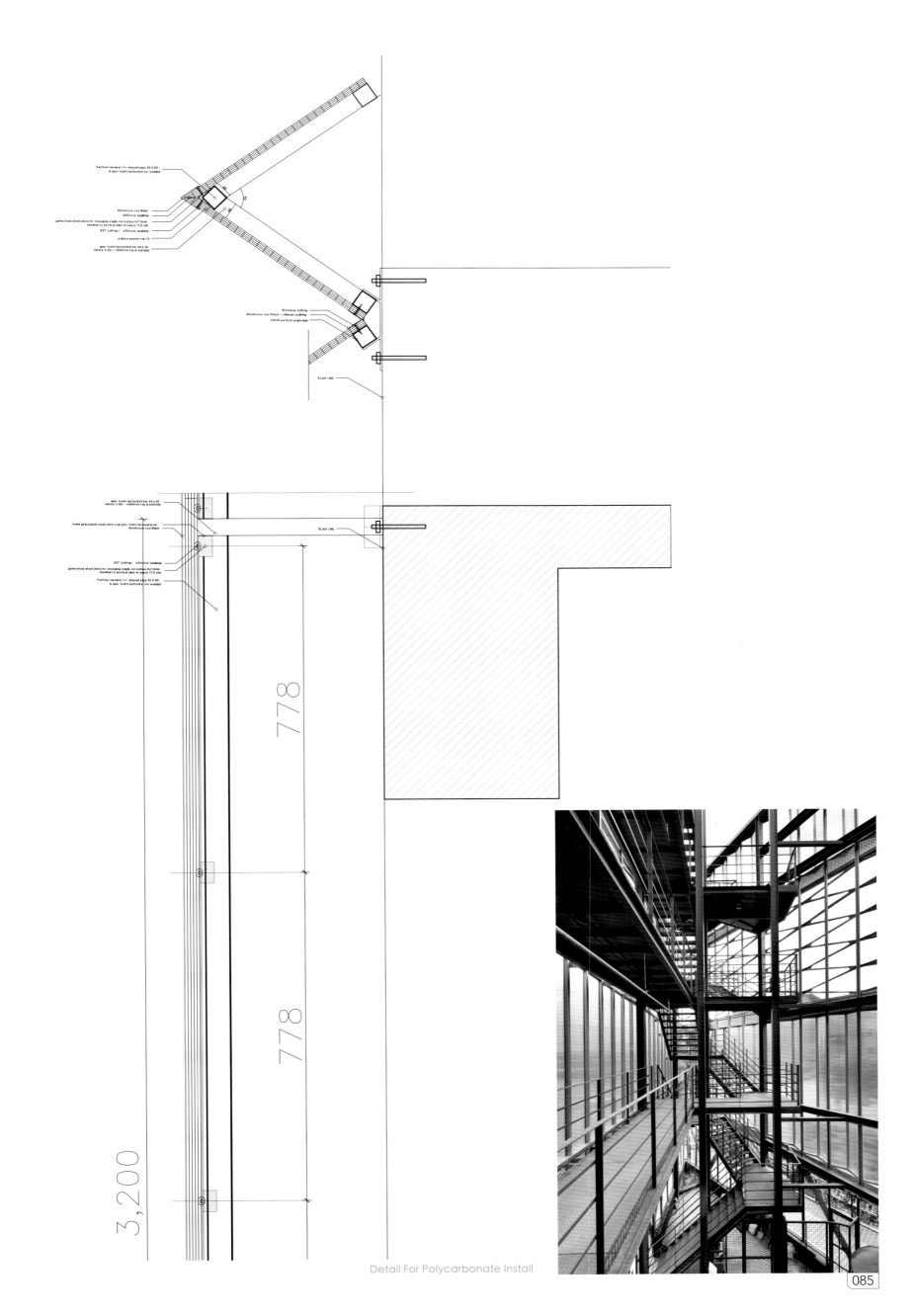

Detail For Polycarbonate Install

Urban Landmark Green Building Environment-friendly Material Slender Mass

城市标志
绿色建筑
环保材料
纤细楼体

Office Building
办公建筑

As the most common architectural form in today's society, office buildings become the symbol to show the development of the world. Many of them have shaped new skylines of the cities due to their unique appearance and great height. However, upon the introduction of the green standard, office designers shift their focus from the height to the comfortableness of the space.

Glass or other transparent materials are usually used for the facade to ensure daylight and thermal effect of the office spaces. Office buildings, especially the high-rise ones are usually designed with prism-like shape or steeple shape, looking slender. In terms of structural design, it pays attention to the combination of functions and comfortableness. Interior decorating materials are selected to be beautiful, elegant and ecological. Urban style and sustainability are the two standards for office building design.

作为经济繁荣、社会进步、科技发展最常见的建筑标志，办公建筑已经成为当代世界运转状况的象征。很多办公建筑因其独特的楼体造型和目不可测的高度而成为城市的天际线。但是，在绿色建筑成为设计师设计的新标准后，办公建筑在突出其高度的同时也更加注重其舒适度，当下，最贵的楼不是最高的办公建筑，而是环境最好、最舒适的办公建筑。

办公建筑在立面上经常采用玻璃或类似玻璃的透明材料来保证办公空间的采光和热能的吸收；在楼体造型上一般采用棱状型或尖塔型，楼体较为纤细，特别是高层办公建筑；在结构设计上注重办公空间的功能性和舒适性相统一；在内部装饰材料的选择上注重大方美观和生态环保相结合。富有浓郁的都市气息和可持续性绿色建筑是设计师设计办公建筑的双重标准。

KEYWORD 关键词	Skin and Facade 表皮与立面	Concrete, Steel Frame 混凝土立面、钢框架立面	
	Materials 材料	Concrete, Glass, Corten Steel 混凝土、塑形玻璃、克尔顿钢	
	Structure 结构	Steel-Frame 钢架结构	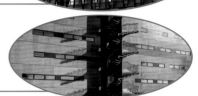

Location: Vannes, France
Architects: Atelier Arcau
Photography: Hervé Le Reste

项目地点：法国瓦纳
建筑设计：法国 Arcau 建筑师事务所
摄影：Hervé Le Reste

Steel Band
Steel Band 办公楼

Features 项目亮点

The steel framework, acting as a filter for the users, provides an ever changing and powerful experience when viewed from "la promenade architecturale" all around Steelband.

钢铁框架的设计保证了一定的私密性，并带来一种不断变化的强劲的视觉体验。

Overview 项目概况

The Steelband project had its origin in 2007 when five firms (auditor, near estate broker, architects, landscape architects and graphic designers) decided to come together to create custom designed sustainable offices. For their purpose, they choose as a starting point the German "Banggruppe" promoting self-development. It took five years to develop and produce the work. Located in the tertiary sector Laroiseau Park, in Vannes, the two buildings, known as "Steelband, are positioned on the fragile fringe between the city and farmland. These two buildings, one two, and the other five storeys, are balanced and interact with one another. Characteristic and powerful, consequently, they become an urban and architectural landmark at the entrance to the town. In the center of the composition, a water garden links interior and exterior spaces. By courtesy to the street and public spaces, car parking is under the buildings, and masked by shear walls, which reinforce the urban line.

该项目是5家公司的办公所在地，是一个可持续发展的办公空间。项目由两个体量组成，一个2层，一个5层，和谐互动。项目位于瓦纳 Laroiseau 公园城乡结合处，一边是城区，另一边是农田。它个性鲜明，理所当然地成为了城镇门户的建筑地标。项目中心的水景园连接了室内外空间。临街的停车场位于地下，剪力墙掩饰，增强都市气息。

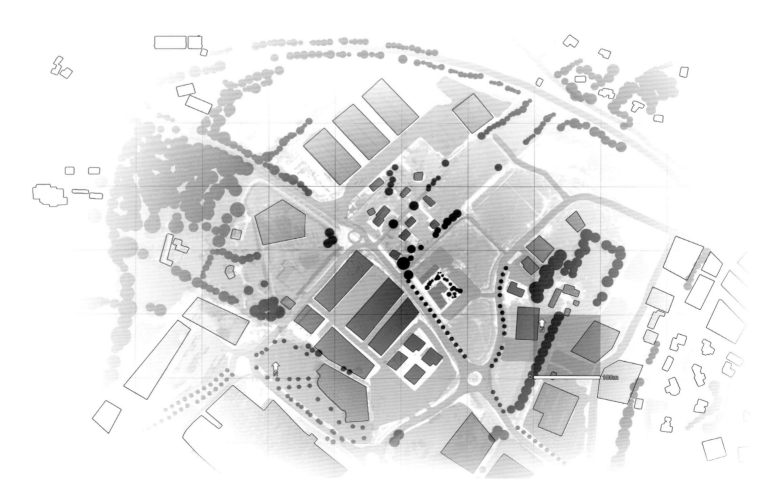

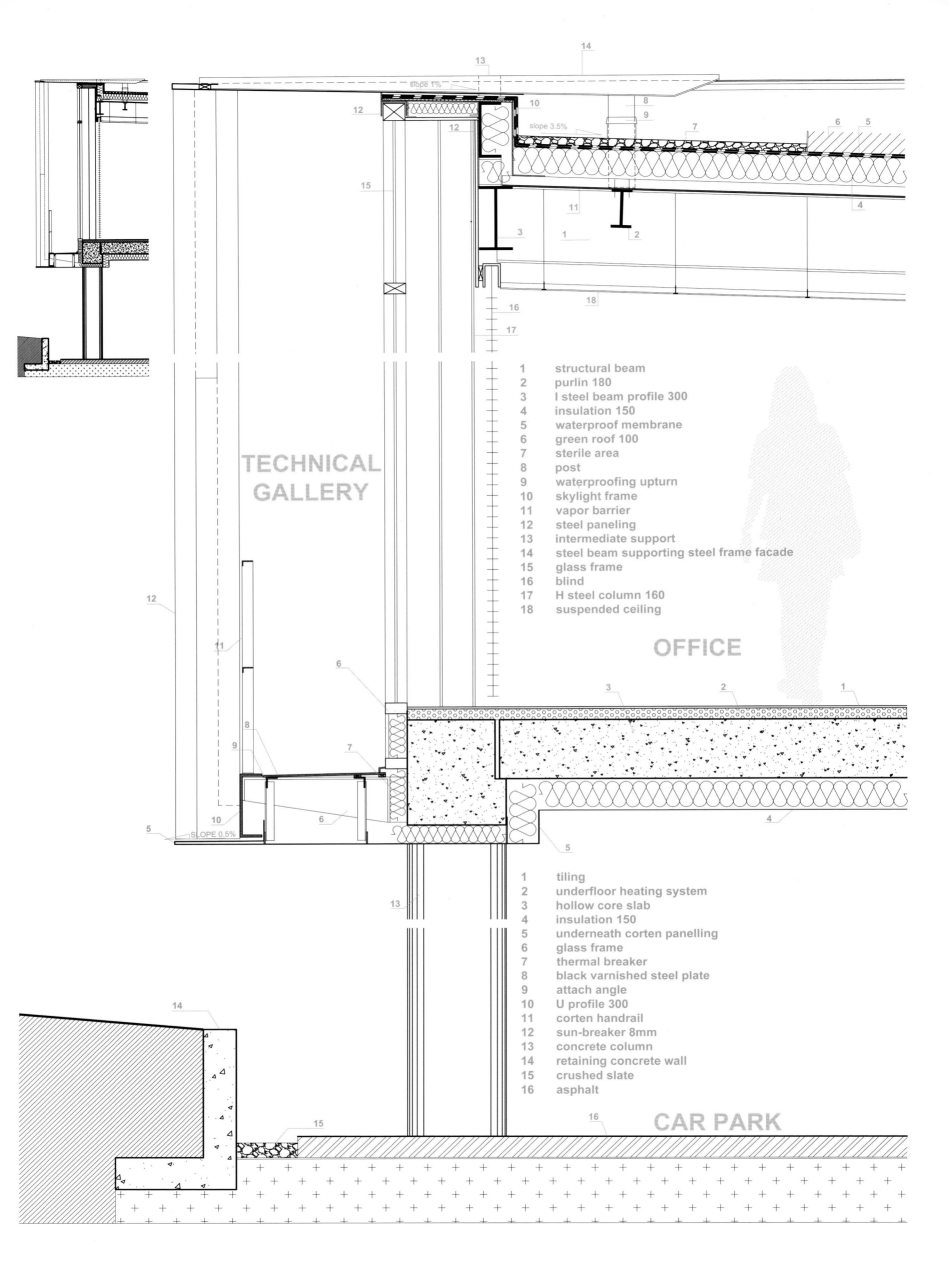

KEYWORD 关键词	Skin and Facade 表皮与立面	White Undulating Facade 白色波状立面
	Materials 材料	Concrete, Aluminum, Wood 混凝土、铝材、木材
	Detail 细部	

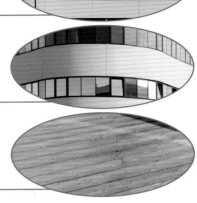

Location: 1 rue Louis Lagorgette – Cenon
Client: SCCV Aalta (Martin Duplantier, Laurent Duplantier, Anouk Debarre)
Architects: Martin Duplantier and Laurent Duplantier architects
Photography: Arthur Pequin

项目地点：法国塞南路易拉戈热特大道1号
客户：SCCV Aalta (Martin Duplantier, Laurent Duplantier, Anouk Debarre)
建筑设计：法国 Martin Duplantier and Laurent Duplantier 建筑师事务所
摄影：Arthur Pequin

Aalta
Aalta 办公楼

Features 项目亮点

The facade has a dynamic urban aspect: undulating through streams, its white shell seems to lead a movement in the neighborhood.

外立面采用白色波状设计，线条流畅，给整个街区带来了动感与活力。

Overview 项目概况

What if we stopped zoning? And what if we projected an office building in the middle of a social housing district? The building Aalta in Cenon takes advantage of different economic incentives and of the development of the Bordeaux tramway to densify and create an active interface between Palmer social district and a quiet neighborhood made of small houses.

项目位于 Palmer 社区和一个由数个小房子组成的安静街区之间，它是经济进步与电车轨道发展的产物。项目地处免税区，功能性多样、灵活性强，因此对一些公司颇具吸引力。

Ground plane

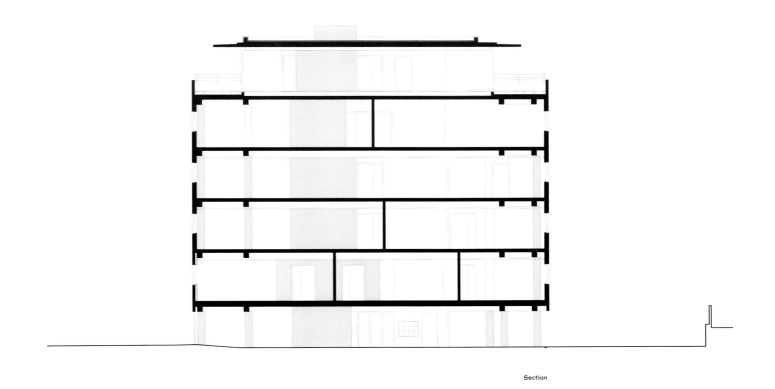
Section

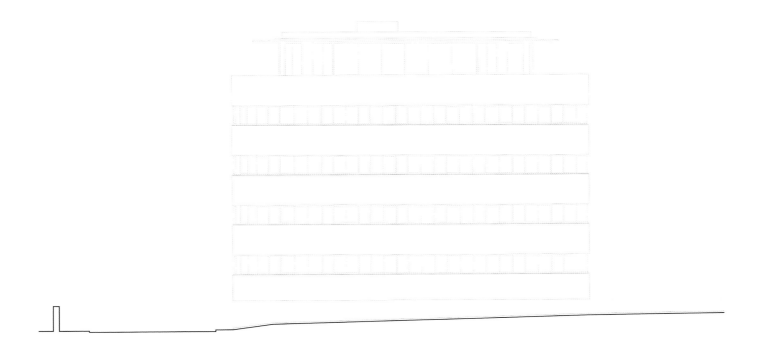
West façade

North façade

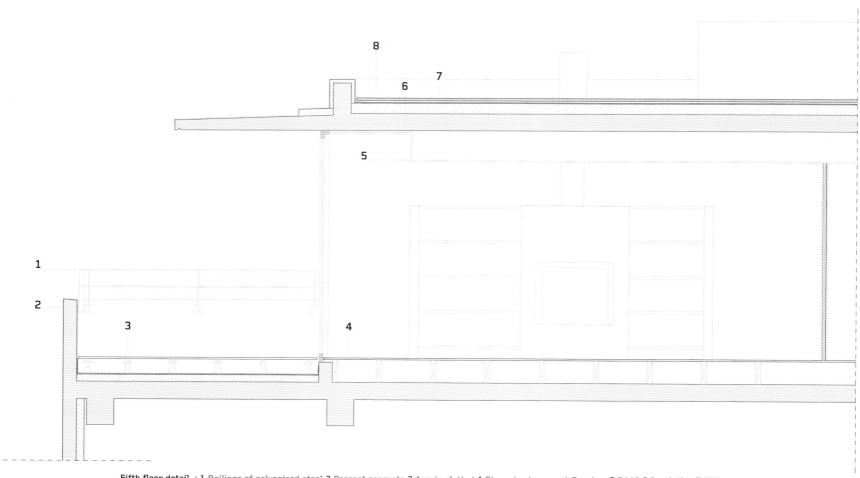

Fifth floor detail : 1 Railings of galvanised steel 2 Precast concrete 3 Acacia slatted 4 Shaped oak parquet flooring 5 BA13 6 Insulation 7 PVC membrane 8 Leaking roof skin

KEYWORD 关键词	Skin and Facade 表皮与立面	Brick Facade 砖材立面
	Materials 材料	Red Brick, Steel Frame, Metal Screens and Glazing 红砖、钢制框架、金属筛网、玻璃
	Detail 细部	

Location: Helsinki, Finland
Client: Ahlström Capital Oy
Architect: Helin & Co Architects
Site Area: 3,300 m²
Floor Area: 14,000 m²

项目地点：芬兰首都赫尔辛基
客户：Ahlström Capital Oy
建筑设计：芬兰 Helin & Co Architects 建筑事务所
占地面积：3 300 m²
建筑面积：14 000 m²

Ahlström Salmisaari Office Building
Ahlström Salmisaari 办公楼

Features 项目亮点

The exterior is dominated by red brick, both in surfaces and as a tectonic structure like the old industrial buildings.

大楼外部材料主要采用红砖及玻璃，使得建筑表面和整体结构看起来更像一个工业建筑。

Overview 项目概况

The office block is located in the Ruoholahti area of Helsinki. Working facilities are universal and modifiable, with emphasis on openness and interaction. Common facilities include a restaurant and a conference centre. Exhibition facilities at the street level enliven the shore promenade. Parking space is provided behind these in the basement as part of a public parking garage.

项目位于芬兰首都赫尔辛基市 Ruoholahti 区。其办公设施可通用也可根据实际情况做调整，强调了其开放性和互动性。公共设施包括餐厅和会议中心。沿街的展览设施给海滨长廊注入了更多的活力。停车空间设置在地下室，作为公共停车场的一部分。

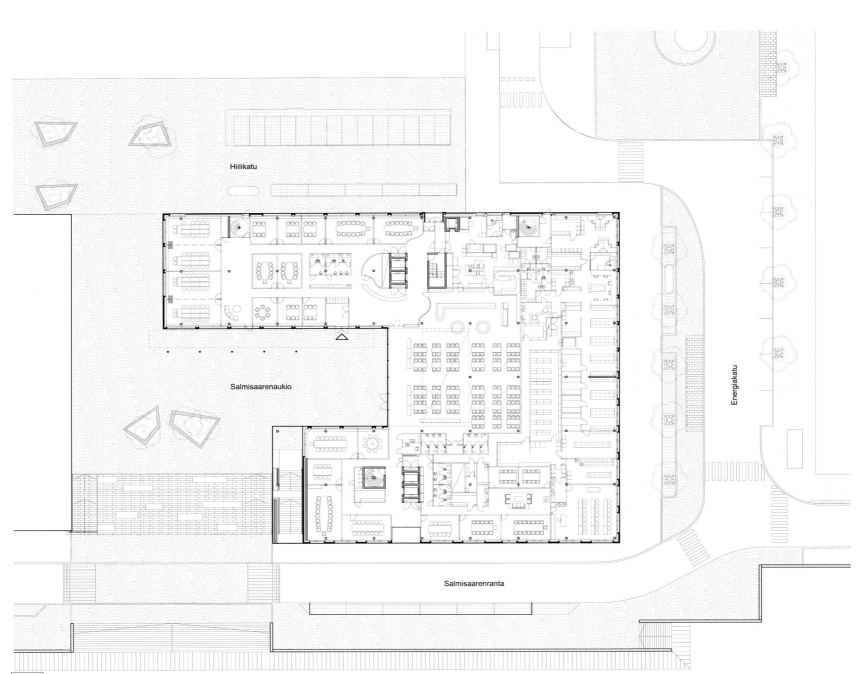

Facade and Materials   立面与材料

Floors wind round a central space that opens up to the market square to the north. The steel frame carries lightweight pre-cast steel units used in the structural sections of the outer walls. The material is most clearly visible in the wind braces of the tall glass wall and in the lattices of the roof lights. Intermediate floors are hollow precast slabs.

The exterior is dominated by red brick, both in surfaces and as a tectonic structure like the old industrial buildings. The thermal load of the large window openings has been compensated with metal screens and special glazing. The site plan and the strict interpretation of it by building control authorities made it necessary to clad the two top floors completely in glass.

项目中心空间面对着市场广场向北延伸，而大厦楼层则沿着此中心空间向上伸展。办公楼的钢制框架支撑着外墙结构所使用的钢筋混凝土部分。该材料用在高大的玻璃幕墙和楼顶灯饰的格子框架上，显得格外引人注目。项目中间的楼层，则采用的是空心预制板。

大楼外部材料主要采用红砖，使得建筑表面和整体结构看起来像一个工业大厦。大型窗户的热负荷由金属筛网和特种玻璃代为抵消。基地的规划需求和建设管理部门的严格规定，使得大楼顶端的两个楼层必须使用玻璃覆盖起来。

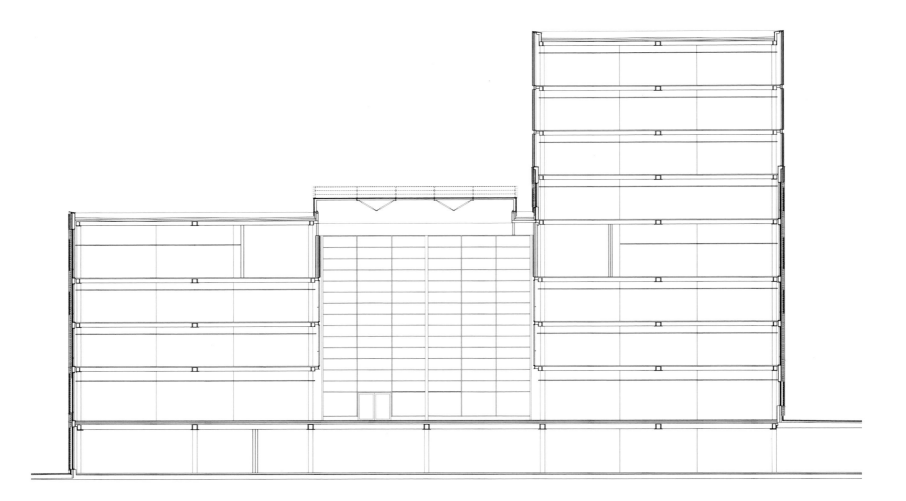

KEYWORD 关键词	Skin and Facade 表皮与立面	Glass Facade 玻璃立面
	Materials 材料	Teflon-coated Fibreglass Mesh 聚乙烯涂层的玻纤网格
	Sustainable Design 可持续性设计	

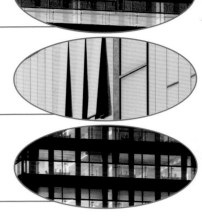

```
Location: Utrecht, The Netherlands
Client: rijksgebouwendienst (government buildings
agency) the hague
Architects: Cepezed Architects
Floor Area: total 53,000 m²
            new construction 23,000 m²
            renovation 27,000 m²
            ground and entrance 3,000 m²
Photographer: Jannes Linders

项目地点：荷兰乌特勒支
客户：荷兰海牙 rijksgebouwendienst
建筑设计：荷兰 Cepezed 建筑师事务所
总建筑面积：53 000 m²
           新建面积 23 000 m²
           改造面积 27 000 m²
           地面和入口面积 3 000 m²
摄影：Jannes Linders
```

Westraven
Westraven 办公楼

Features 项目亮点

The ensemble has a great variety of sustainability-related aspects, such as the second-skin façade and climate ceilings as well as the underground warmth and cold storage. A series of integrated design solutions that are not only aesthetic and functional, but are useful in terms of material saving as well.

整座建筑从多方面体现了可持续的设计理念，如表皮涂层和气候调节天花板，地下储热和降温系统等，不仅美观实用，也有效地节省了材料和能源。

Overview 项目概况

For a lengthy period, the Westraven office building in Utrecht was a heavy, closed and stuffy colossus that was on the verge of being demolished. The eventually preferred revitalization aimed at reflecting the modern values of user Rijkswaterstaat (department of Public Works): openness, transparency, professionalism and sustainability. The proposals advanced by Cepezed harmonized best with this goal.

The complex comprises existing and new construction is located close to where the A12 motorway crosses the Amsterdam-Rijn canal. The programme includes office space, conference facilities, a national meeting centre, a communications centre, and the Rijkswaterstaat Future Center. More than 2,000 people work in the complex.

在过去很长一段时间内，位于乌特勒支的 Westraven 办公楼给人的感觉便是一座沉重、古板而又封闭的庞然大物，处在被拆除的临界状态。最终选定的翻新方案旨在突出其使用者——公共工程部的当代价值：开放、透明、专业和持续性。由 Cepezed 提出的这项方案与该目标完美契合。

该工程是一个新旧建筑结合的综合型大型项目，位于阿姆斯特丹——莱茵运河与 A12 高速公路在乌特勒支汇合处之间的 Westraven 地区。设计内容包括办公空间、会议设施、国家会议中心、通信中心、公共工程部的未来中心。这里可以容纳 2 000 以上的人员办公。

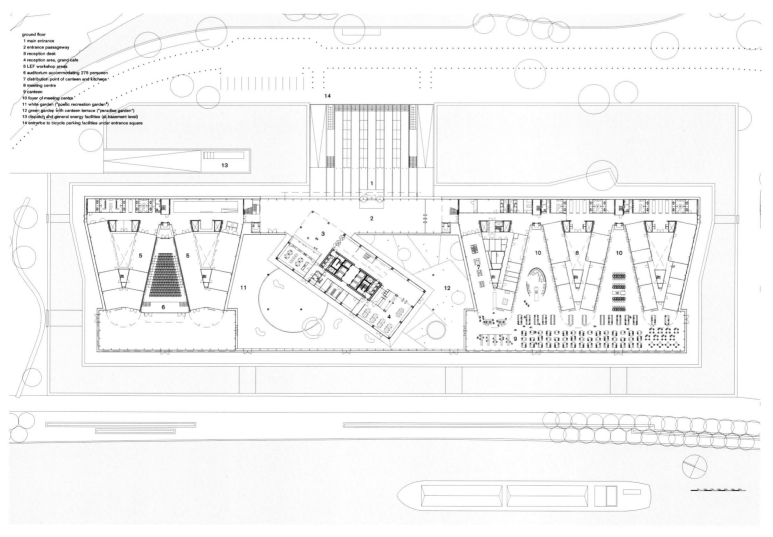

ground floor
1 main entrance
2 entrance passageway
3 reception desk
4 reception area, grand café
5 LEF workshop areas
6 auditorium accommodating 275 personen
7 distribution point of canteen and kitchens
8 meeting centre
9 canteen
10 foyer of meeting centre
11 white garden ("poetic recreation garden")
12 green garden with canteen terrace ("paradise garden")
13 dispatch and general energy facilities (at basement level)
14 entrance to bicycle parking facilities under entrance square

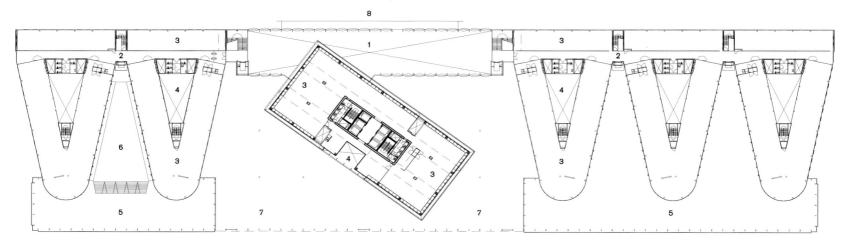

ground plan of third floor
1 entrance passage
2 main entrance/ aorta
3 office
4 void
5 conservatory
6 auditorium roof
7 inner garden (sheltered)
8 artwork

Skin and Facade 表皮与立面

The new façades are largely made of glass to allow a panoramic view and abundant incidence of daylight. Sections that can be opened enable natural ventilation. A second-skin façade of mainly teflon-coated fibreglass mesh prevents wind nuisance, functions as a sunbreak, leaves the view intact, and also forms an aesthetic component that imposes a strongly horizontal articulation. The renewed climate system consists largely of project-specific climate ceilings between the concrete floor beams.

设计大面积地采用玻璃立面，以获取最大化的景观和光照，同时也利于自然通风。建筑表皮主要采用聚乙烯涂层的玻纤网格，使建筑免受风力和日照损害，既保留了窗外景色，又打造了富有美感的建筑形体。更新的气候系统在水泥楼板梁之间采用了特制的气候天花板。

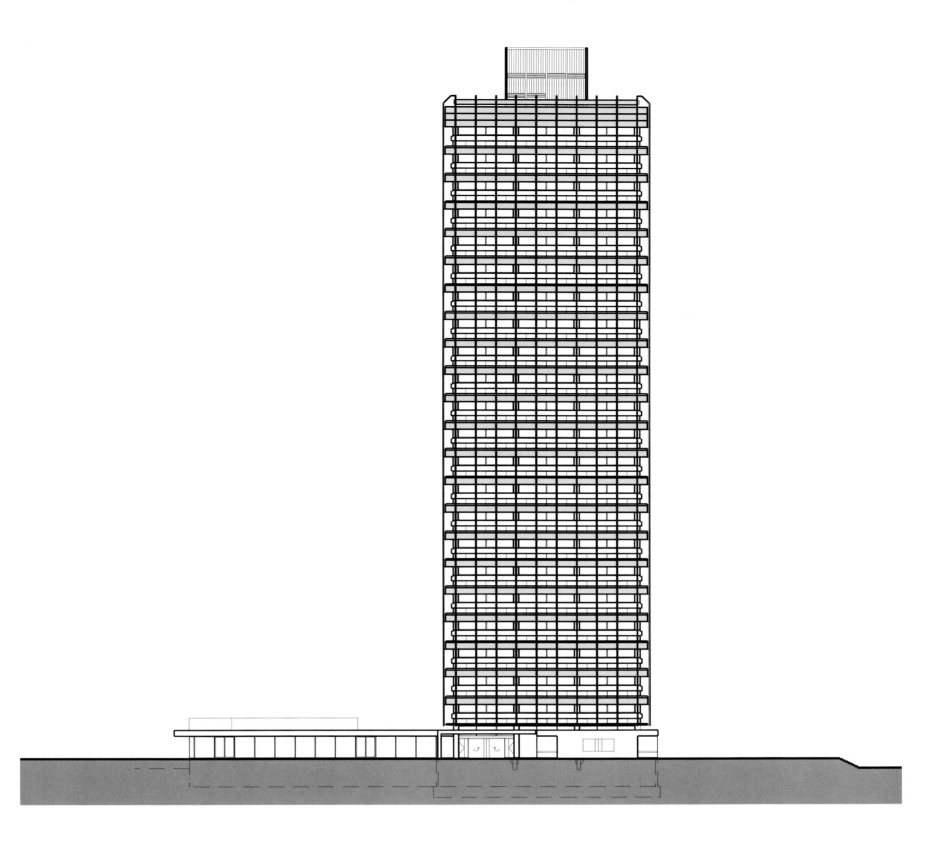
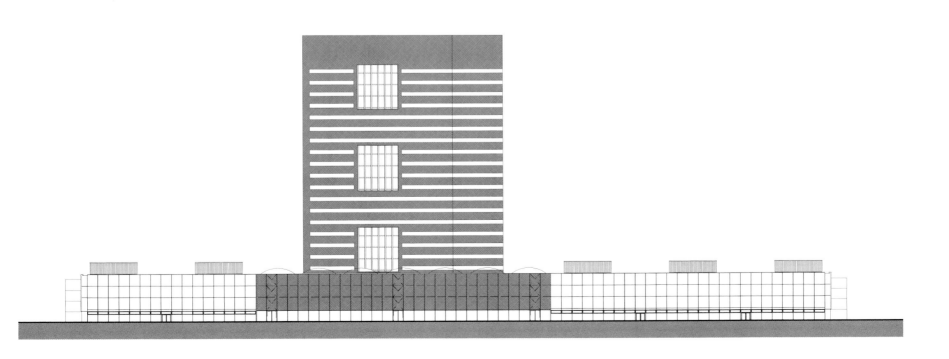

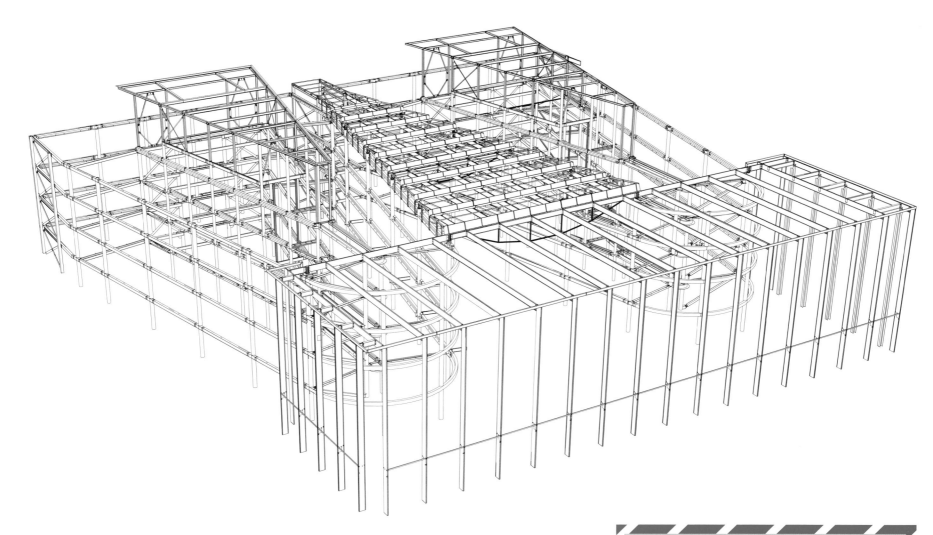

Longtidunal cross-section of the high-rise block
1. reception desk and reception area
2. meeting centre
3. technical services floors
4. offices
5. void
6. services areas for air-conditioning
7. storage and archives

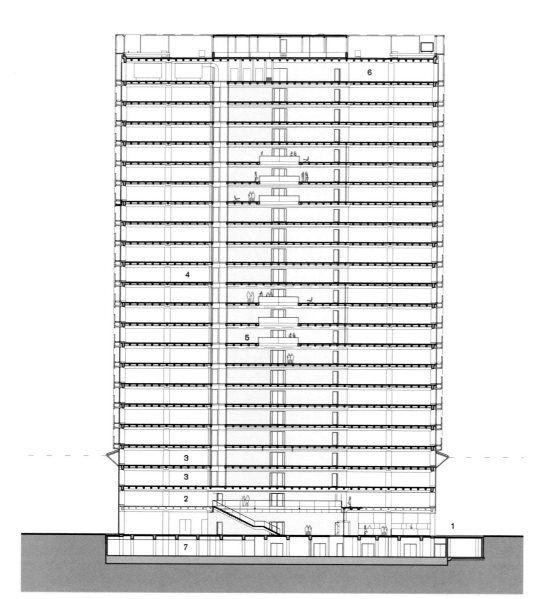

Structure and Shape
结构与造型

The high-rise section, accommodating more than 27,000 m², was stripped down to its concrete skeleton and equipped with five enormous voids that introduced much daylight and spatial experience into the building, and also contribute to good orientation.

At the foot of the high-rise, an extension measuring 23,000 m² contains five triangular office wings that offer a view of the adjacent canal to as many rooms as possible. The triangles are linked at the basis by means of an efficient traffic aorta and at the rounded tips by a large, encompassing conservatory in which the spaces between the wings have been roofed over with transparent material. With a water basin surrounding it, the complex in embedded in a park-like environment of around 3,000 m².

建筑面积超过 27 000 m² 的现有高层建筑已经完全被整修过并重新组织，之前的混凝土结构被拆除，形成五个巨型空间，使建筑享有更多光照，同时丰富了内部的空间体验，而且对工作人员和访客确定方向也非常重要。

高层建筑的底部进行了扩建，扩建面积达 23 000 m²，包括五座三角形办公翼楼。其造型设计使建筑内部尽可能多地欣赏到附近的运河景观。翼楼底部通过高效的交通流线相连，而其圆形顶端则由一座温室围合，并设有透明屋顶。建筑被水围合，如同镶嵌在 3 000 m² 的公园般的环境中。

the building prior to renovation and
new construction
above
typical office storey

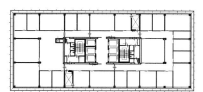

below
ground floor

1 entrance and receeption desk
2 dispatch
3 reprography
4 kitchen
5 canteen
6 meeting centre
7 patio

west façade prior to renovation and
new construction

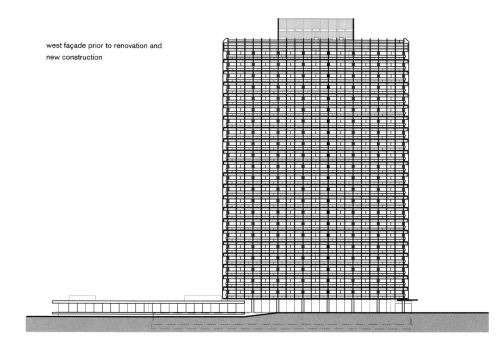

south-west façade gevel (canal side)
1 bicycle parking facilities
2 dispatch and "energy building"

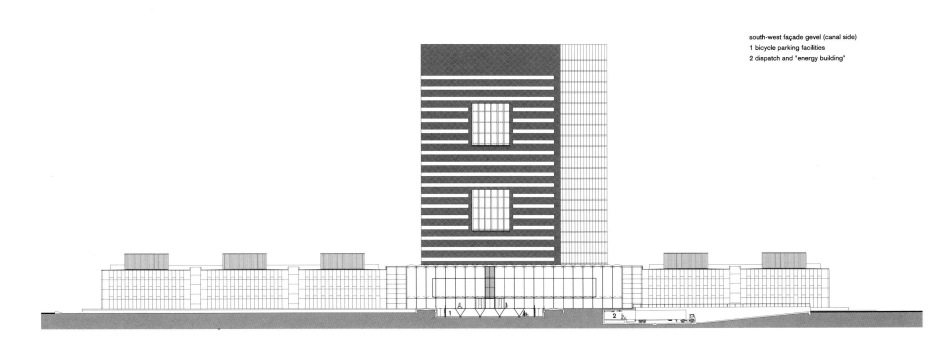

section
1 reception area
2 meeting centre
3 general and technical service departments
4 offices
5 void
6 installations
7 storage and archives
8 auditorium
9 LEF "workshop" areas
10 kitchen and distribution point
11 sheltered inner garden

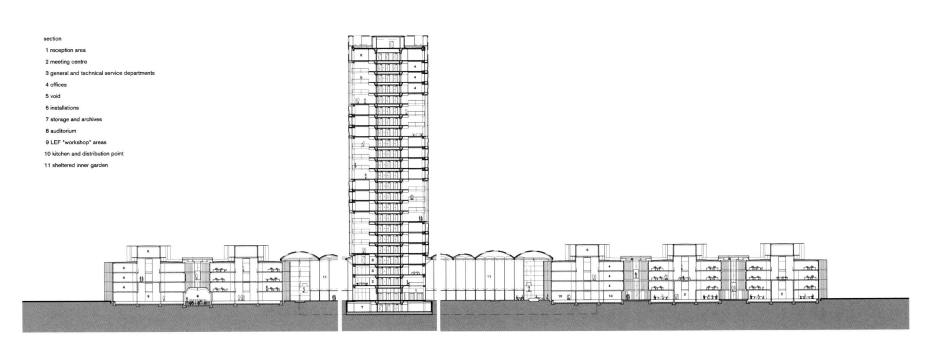

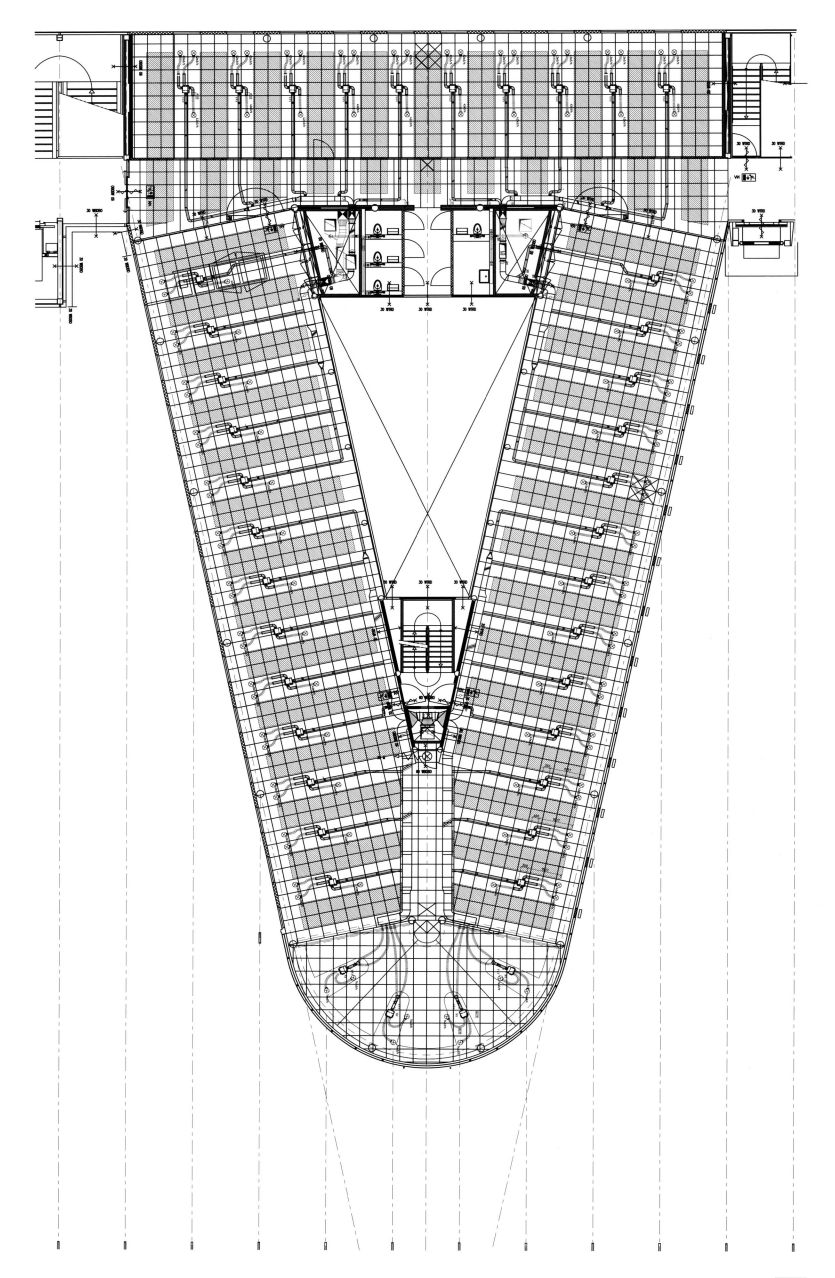

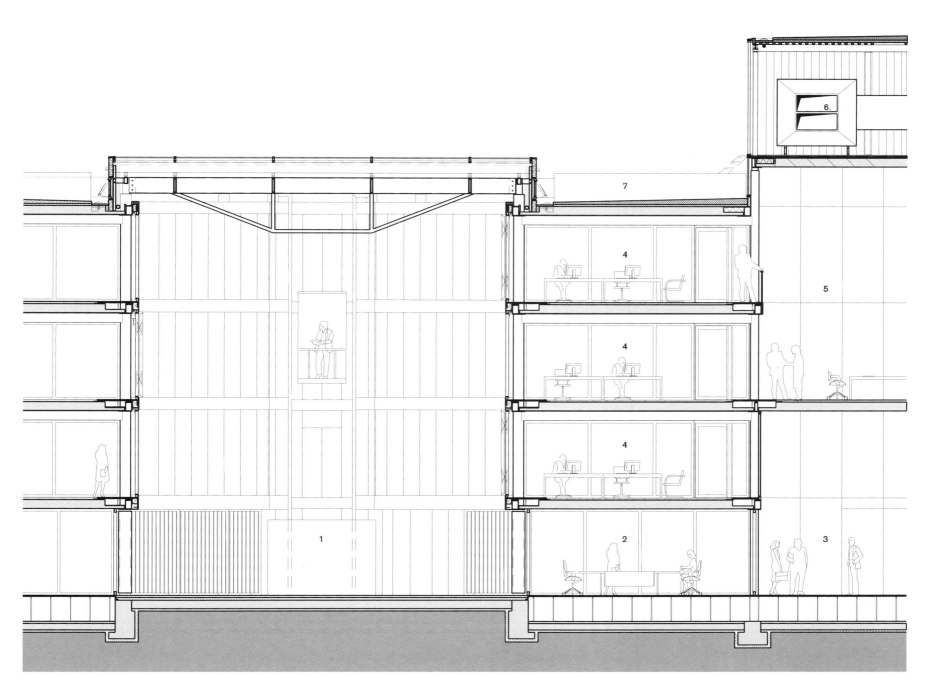

Sustainable Design   可持续性设计

The ensemble has a great variety of sustainability-related aspects. In addition to the second-skin façade and climate ceilings, there are also underground warmth and cold storage, concrete core activation in the new-construction floors, a division into various climate zones, and a series of integrated design solutions that are not only aesthetic and functional, but are useful in terms of material saving as well.

整座建筑从多个方面体现了可持续的设计理念。除了表皮涂层和气候调节天花板，设计还在新的建筑楼层中采用了地下储热和降温系统、混凝土芯激活系统等，针对不同类型的空间进行不同的气候调节。这些综合设计不仅起到了美化和实用的效果，同时也有效地节省了材料和能源。

structure of façade
between offices and conservatory
(inner façade)

left

1 wingfloor equipped with Velta net for concrete core activation
2 steel box column, filled with concrete with steel flange beams
3 hollow in floors with tracks for technical installations, with glass-fibre anhydrite tiles above
4 construction baffle for air overflow with perforated stainless-steel panel to benefit the acoustics in the conservatory
5 two-scale glass modular office wall
6 parallel top-hung window equipped with perforated stainless-steel plating for the acoustics in the conservatory
7 compound steel box beams 80 x 500 mm
8 sun-resitant insulation glass on insulating steel gutters, with glasshouse fabric underneath
9 ventilation opening
10 rockwool insulation and PVC roof covering
11 rotating privacy strip and acoustic muffling, anodized (colourless) and perforated aluminium
12 insulated hollow-core plate and under-floor heating, with a finishing of oak parquet
13 floor hollow for ventilation channels and pipes and ducts for meeting wings
14 conservatory lighting in floor periphery

right

1 conservatory
2 meeting areas
3 foyer meeting centre
4 offices
5 void
6 services areas for air conditioning
7 return route for air from conservatory

Section

1. meeting centre foyer
2. main entrance with raised floor tiles endowed with stainless-steel finishing
3. sanitairy room with Corian washbasin
4. main route for electricity and data cabling and sprinkler installations
5. sanitary groups and shafts for main supply
6. offices
7. void with meeting areas
8. technical service area with air-conditioning units
9. sun-resistant insulating glass
10. stainless-steel unit for coffee machines

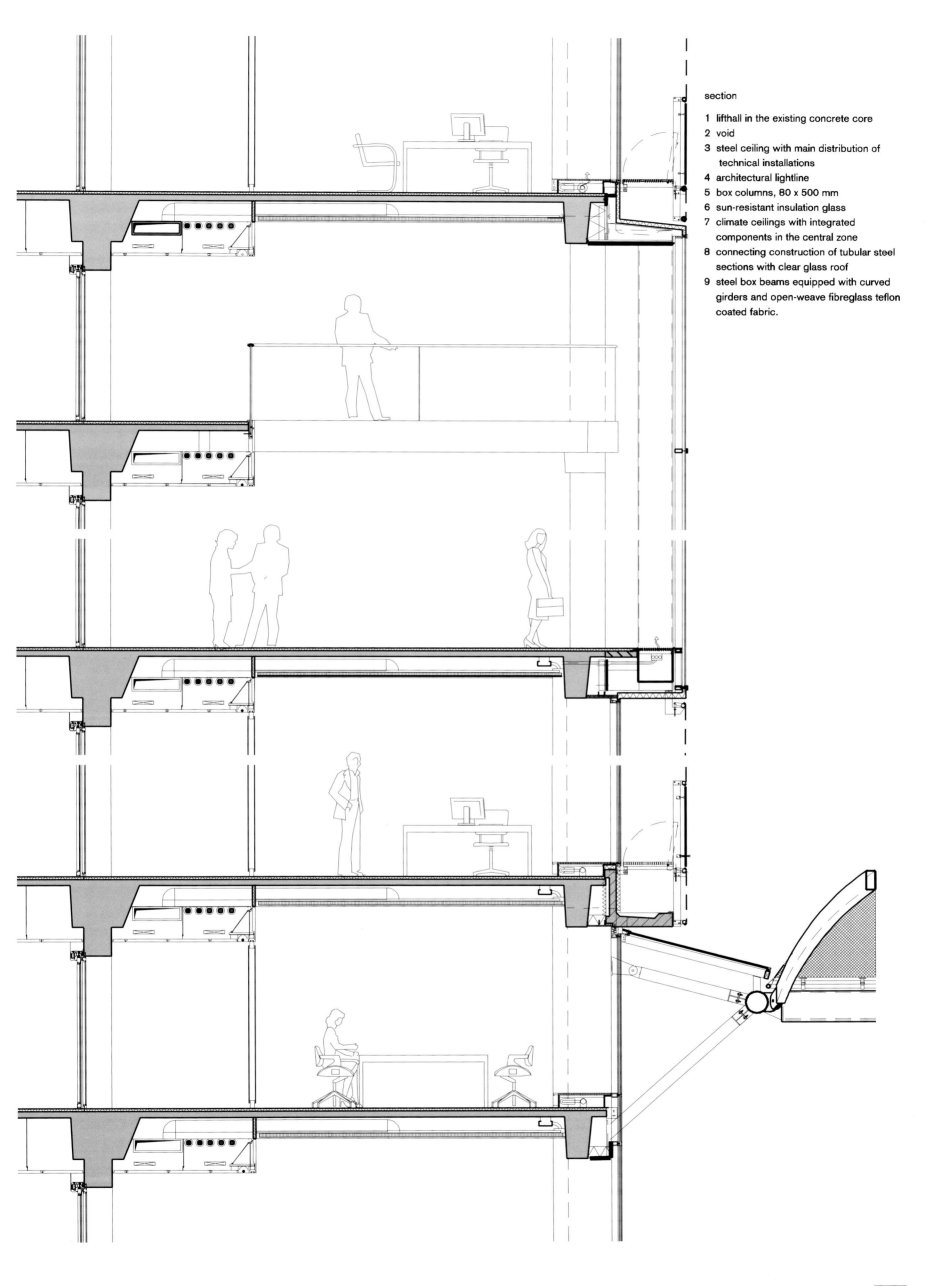

section

1. lifthall in the existing concrete core
2. void
3. steel ceiling with main distribution of technical installations
4. architectural lightline
5. box columns, 80 x 500 mm
6. sun-resistant insulation glass
7. climate ceilings with integrated components in the central zone
8. connecting construction of tubular steel sections with clear glass roof
9. steel box beams equipped with curved girders and open-weave fibreglass teflon coated fabric.

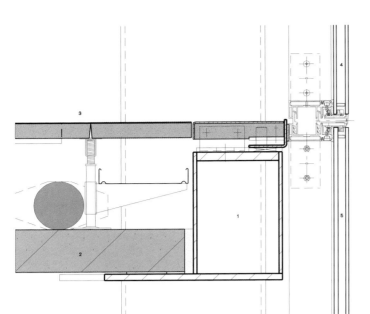

Exterior façade of offices:
detail of floor edging and horizontal detail

1. steel single-flange beam, with fire-resistant coating
2. wing floor equipped with piping for concrete core activation
3. raised floor tiles, fibreglass-reinforced anhydrite
4. sun-resistant insulating glass, shatter-proof
5. top-hung sash window with sun-resistant insulating glass
6. sandwich panel of stainless steel, rock wool, white-coated steel plating
7. rubber seal

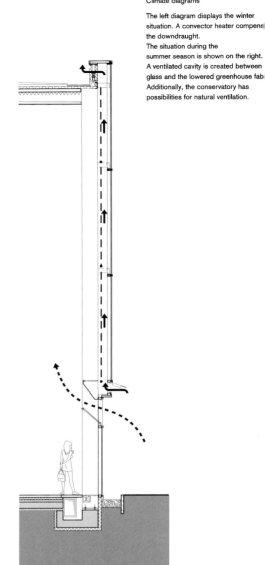

Climate diagrams

The left diagram displays the winter situation. A convector heater compensates the downdraught.
The situation during the summer season is shown on the right. A ventilated cavity is created between glass and the lowered greenhouse fabric. Additionally, the conservatory has possibilities for natural ventilation.

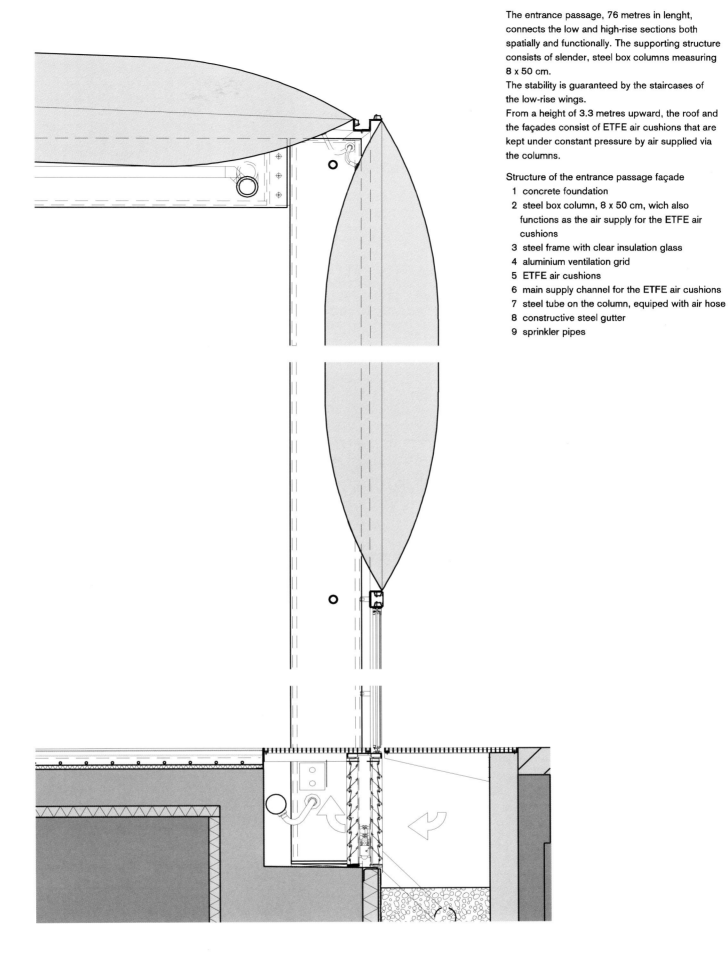

The entrance passage, 76 metres in lenght, connects the low and high-rise sections both spatially and functionally. The supporting structure consists of slender, steel box columns measuring 8 x 50 cm.
The stability is guaranteed by the staircases of the low-rise wings.
From a height of 3.3 metres upward, the roof and the façades consist of ETFE air cushions that are kept under constant pressure by air supplied via the columns.

Structure of the entrance passage façade
1. concrete foundation
2. steel box column, 8 x 50 cm, wich also functions as the air supply for the ETFE air cushions
3. steel frame with clear insulation glass
4. aluminium ventilation grid
5. ETFE air cushions
6. main supply channel for the ETFE air cushions
7. steel tube on the column, equiped with air hose
8. constructive steel gutter
9. sprinkler pipes

structure of conservatory façade
1. concrete foundation beam
2. steel box beam 8 x 50 cm
3. aluminium modular façade with insulating and sun-resistant glass
4. steel roof with a ceiling of perforated, clear-anodized aluminium between the joists
5. sun-resistant glasshouse fabric with tubular motor
6. air supply for cavity ventilation
7. air discharge for cavity ventilation
8. lightning and convector unit, supply of water and electricity takes place via steel column
9. convector and main supply track

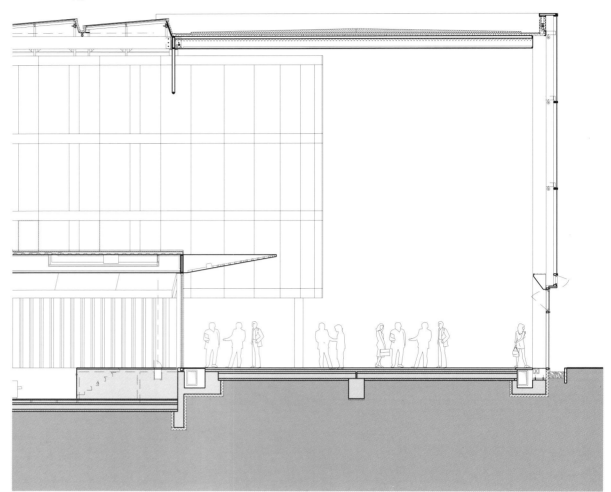

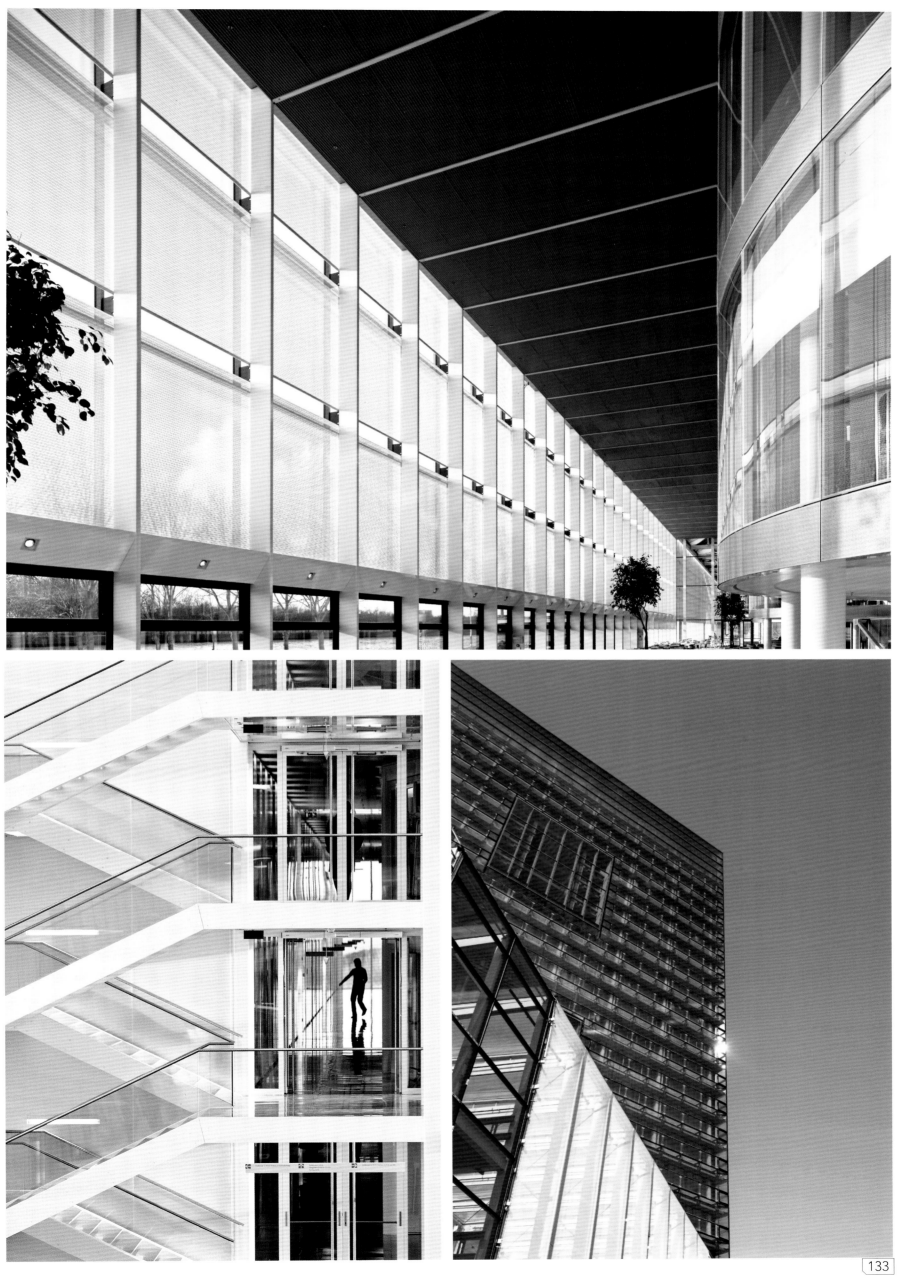

KEYWORD 关键词	Skin and Facade 表皮与立面	Glass Façade 玻璃立面
	Materials 材料	Glass, Limestone, Black Granite, Aluminium Alloy 玻璃、石灰岩、黑花岗岩、铝合金
	Architectural Detail 建筑细部	

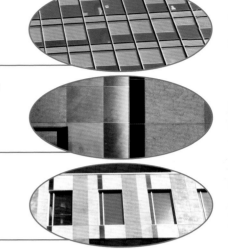

Location: Quebec, Canada
Architectural Design: Cardinal Hardy & Le Groupe Arcop

项目地点：加拿大魁北克
建筑设计：Cardinal Hardy 建筑设计事务所，Arcop 设计集团

Head office of Quebecor
魁北克总部

Features 项目亮点

The materiality of the new building is in continuity with the existing building, endowing the grouping with homogeneity of color and texture, thus create a neutral-coloured link between neighboring buildings with their variety of materials.

建筑材料的选择与旧建筑一致，颜色与纹理也体现了一致性，从而建立起与临近建筑一致的色彩联系。

Overview 项目概况

Using the concepts of recycling and integration to achieve this project, three conditions seemed essential to the architects: the concept had to reconnect the building to the urban context and environment in every sense; the team had to adopt refined approaches to urban design and project management; and the project had to be accomplished based on the quality of the design and the materials.

该项目的设计理念是循环和融合，三个关键条件是：设计理念结合城市环境，设计团队需采用精确的方法来设计和管理项目，项目建成需建立在设计质量及材料的选择上。

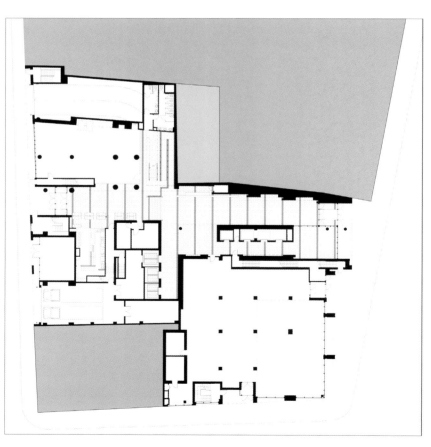
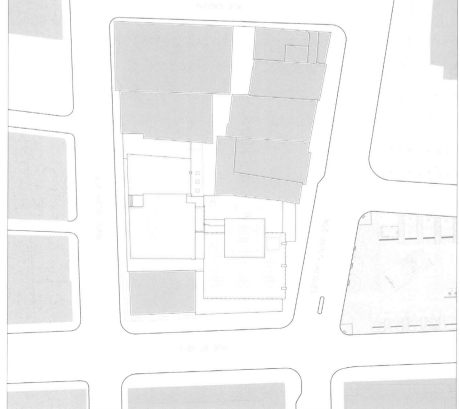

Facade and Materials 立面与材料

The idea of making full use of the structure of the existing building by grafting a second building in continuity with it was the beginning of a long series of principles that led the architects to think about their project around the concepts of recycling and reinterpretation of the existing architectural elements. On the ground floor, this idea is conveyed by the creation of a new entrance hall that takes advantage of the relocation of the old parking-access ramp. Thus, a new two-storey-high public passageway crosses the block from one side to the other, connecting Rue Notre-Dame to Victoria Square. On the upper storeys, levels of circulation are created connecting the new circulation levels with the older ones by going through the black masonry wall on the south side, which has been opened up. This link is conveyed by a living and elegant architectural detail: completely glazed passageways on fourteen storeys that interconnect the volume of ebonite black masonry, in which the new building's elevators are hidden, with the halls and stairways of the existing building. While the light reflects on the glazed surface of the linking nodule, the low-emissivity film on the glass reflects a rose-violet light onto the party-wall masonry, which warms the entire area. On the last seven storeys, a stairway rises above the glass volume to appear cantilevered over the new black-masonry wall, animating the vertical circulations on the main façade of the building.

The materiality of the new building is in continuity with the existing building, endowing the grouping with homogeneity of color and texture. The uniformity of materials responds to a concern for the project's urban integration by deliberately adopting a scale and typology similar to those of adjoining buildings, while having a language on the façade that is both contemporary and simple and highlights that of the old architecture through a play of contrasts. Limestone, black granite, glass, and aluminum are used to create a neutral-coloured link between neighboring buildings with their variety of materials.

通过将新建筑连续地移植到旧建筑上，并充分利用旧建筑的框架是一系列设计原则的前提条件，使设计师从循环利用和重新诠释现存设计元素的角度考虑设计方案。一层大厅的入口设计体现了这一想法，入口设计通过旧停车场的入口通道的重置从而创造了一个从建筑一侧到另一侧的两层楼高的公共通道，连通着圣母街到维多利亚广场。高层楼设计中，通过南部砌筑墙的连通可使水平方向的新旧通道连接起来。这种联系通过生动而又优雅的建筑细节体现出来：14层光滑的走廊与硬橡胶的黑色砌体相互连接从而将电梯隐藏起来，突出走廊与楼梯的存在感。灯光反射到玻璃立面连接的结节处，低辐射镀膜玻璃将紫玫瑰色的灯光反射到砌筑墙上，温暖了整个区域。最高的那七层建筑中，其楼梯呈悬臂式，位于黑色砌筑墙上，使建筑外立面的垂直流线系统更显生动。

建筑材料的选择与旧建筑一致，颜色与纹理也体现了一致性。材料的一致性通过刻意采用与邻近建筑相似的规模及类型来呼应城市融合，外立面的建筑语言即体现了建筑现代、简洁的特点，突出该项目与旧建筑的对比性。石灰岩、黑花岗岩、玻璃以及铝合金等各类建筑材料的使用打造了与临近建筑一致的色彩联系。

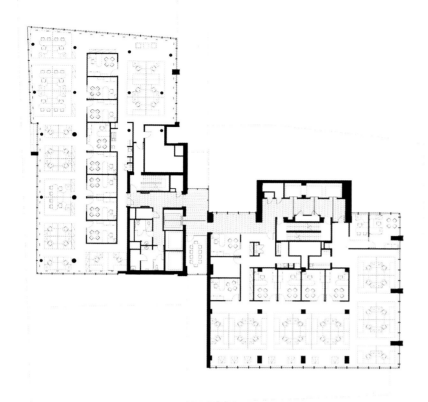
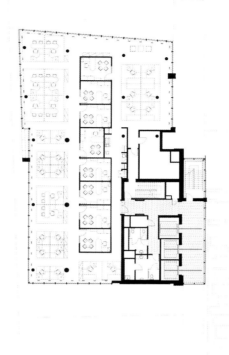
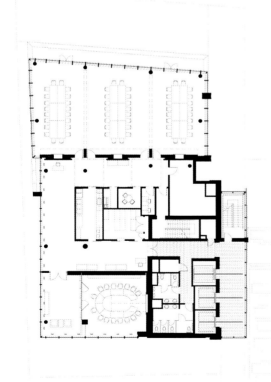
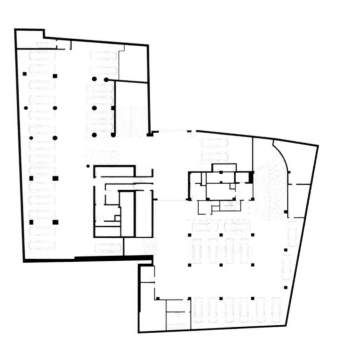

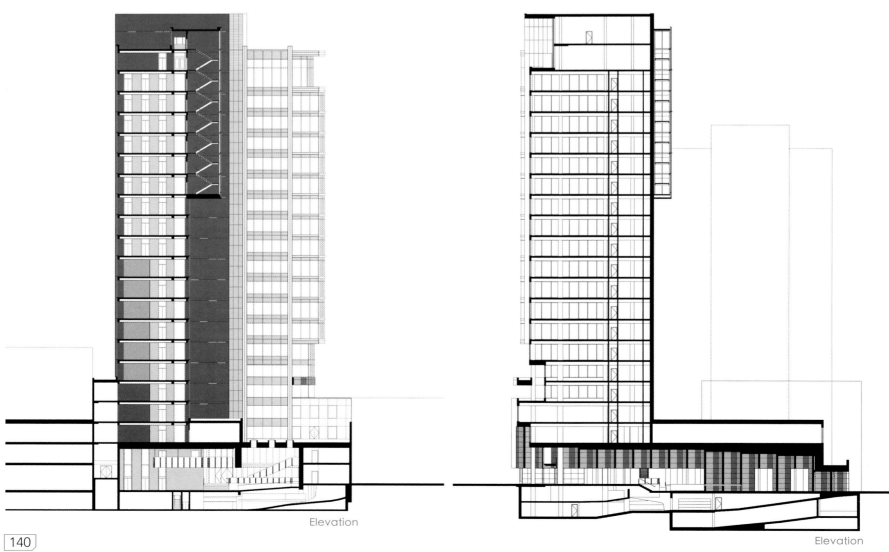

Elevation

Elevation

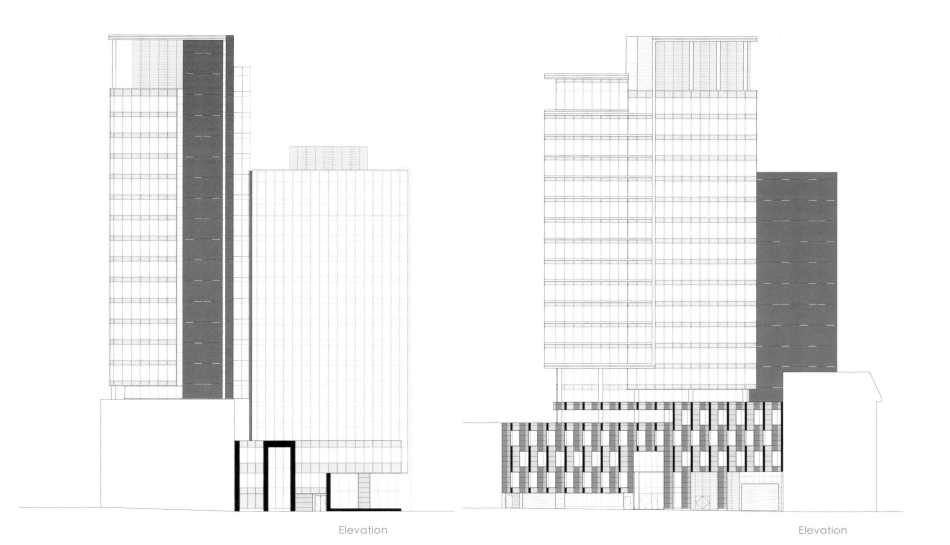

Elevation

Elevation

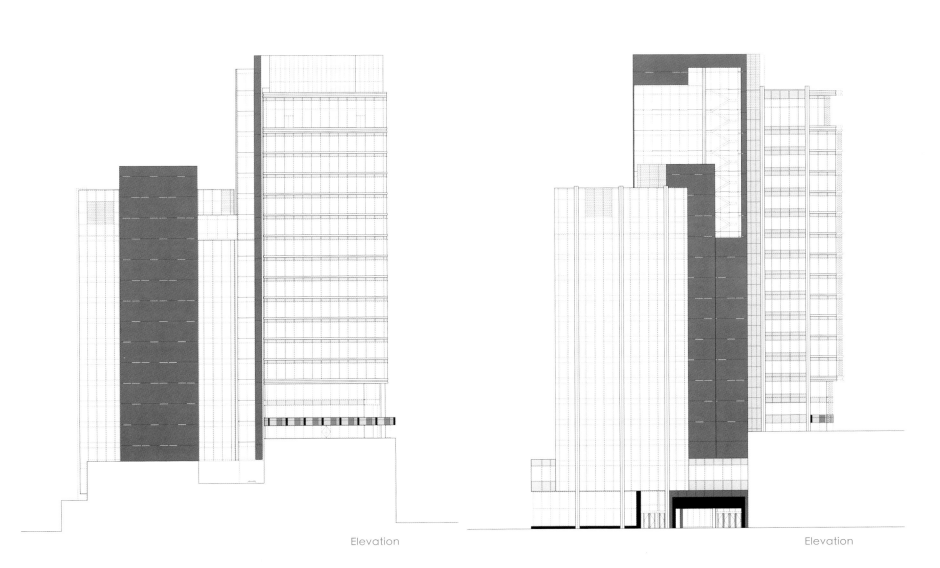

Elevation

Elevation

KEYWORD 关键词

Skin and Facade 表皮与立面 | Mixed Façade 混合立面

Materials 材料 | Glass 玻璃

Sustainable Design 可持续性设计

Location: Winnipeg, Manitoba, Canada
Client: Manitoba Hydro
Architectural Design: Kuwabara Payne McKenna Blumberg Architects
Associate Architects: Smith Carter Architects and Engineers
Floor Area: 64,500 m²

项目地点：加拿大马尼托巴温尼伯
客户：马尼托巴水电局
建筑设计：加拿大 KPMB 建筑设计事务所
合作设计：Smith Carter 设计事务所
建筑面积：64 500 m²

Manitoba Hydro Place
马尼托巴水电局大楼

Features 项目亮点

The building relies on passive free energy, such as materials, orientation and other characteristics provide most energy for cooling and heating, thus avoid affected by extreme climate.

该建筑主要通过被动方式来节能，例如材料、朝向等特点，为建筑制冷和采暖提供大部分能源，使建筑远离了极端的气候反应。

Overview 项目概况

Manitoba Hydro is the major energy utility in the Province of Manitoba, the fourth largest energy utility in Canada and offers some of the lowest electricity rates in the world. Manitoba Hydro Place is the first of the next generation of sustainable buildings integrating time-tested environmental concepts in conjunction with advanced technologies to achieve a "living building" that dynamically responds to the local climate. Located in downtown Winnipeg, the city is known for its extreme climate. The 64,500 m² tower is targeting less than 100 kWh/m²/a compared to 400 kWh/m²/a for a typical large scale North American office tower located in a more temperate climate.

　　Manitoba Hydro 大厦，作为加拿大的第四大公用能源事业公司新的总部，将成为加拿大最可持续发展的建筑。大厦坐落在温尼伯市（拥有北美极端气候的城市之一）中心，该面积为 64 500 m² 的大厦在节能方面积极的立场得到了高度的赞扬，预计为上一个典型办公楼耗能的四分之一，使建筑远离了极端的气候反应。

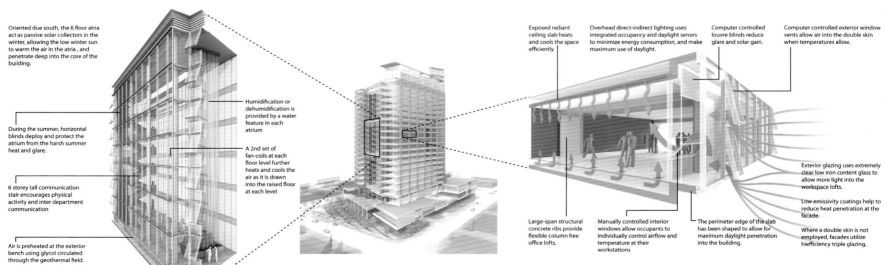

Oriented due south, the 6 floor atria act as passive solar collectors in the winter, allowing the low winter sun to warm the air in the atria, and penetrate deep into the core of the building.

During the summer, horizontal blinds deploy and protect the atrium from the harsh summer heat and glare.

6 storey tall communication stair encourages physical activity and inter department communication

Air is preheated at the exterior bench using glycol circulated through the geothermal field.

Humidification or dehumidification is provided by a water feature in each atrium

A 2nd set of fan-coils at each floor level further heats and cools the air as it is drawn into the raised floor at each level

Large-span structural concrete ribs provide flexible column free office lofts.

Exposed radiant ceiling slab heats and cools the space efficiently

Overhead direct-indirect lighting uses integrated occupancy and daylight senors to minimize energy consumption, and make maximum use of daylight.

Manually controlled interior windows allow occupants to individually control airflow and temperature at their workstations

Computer controlled louvre blinds reduce glare and solar gain.

The perimeter edge of the slab has been shaped to allow for maximum daylight penetration into the building.

Computer controlled exterior window vents allow air into the double skin when temperatures allow.

Exterior glazing uses extremely clear low iron content glass to allow more light into the workspace lofts.

Low emissivity coatings help to reduce heat penetration at the facade.

Where a double skin is not employed, facades utilize hiefficiency triple glazing.

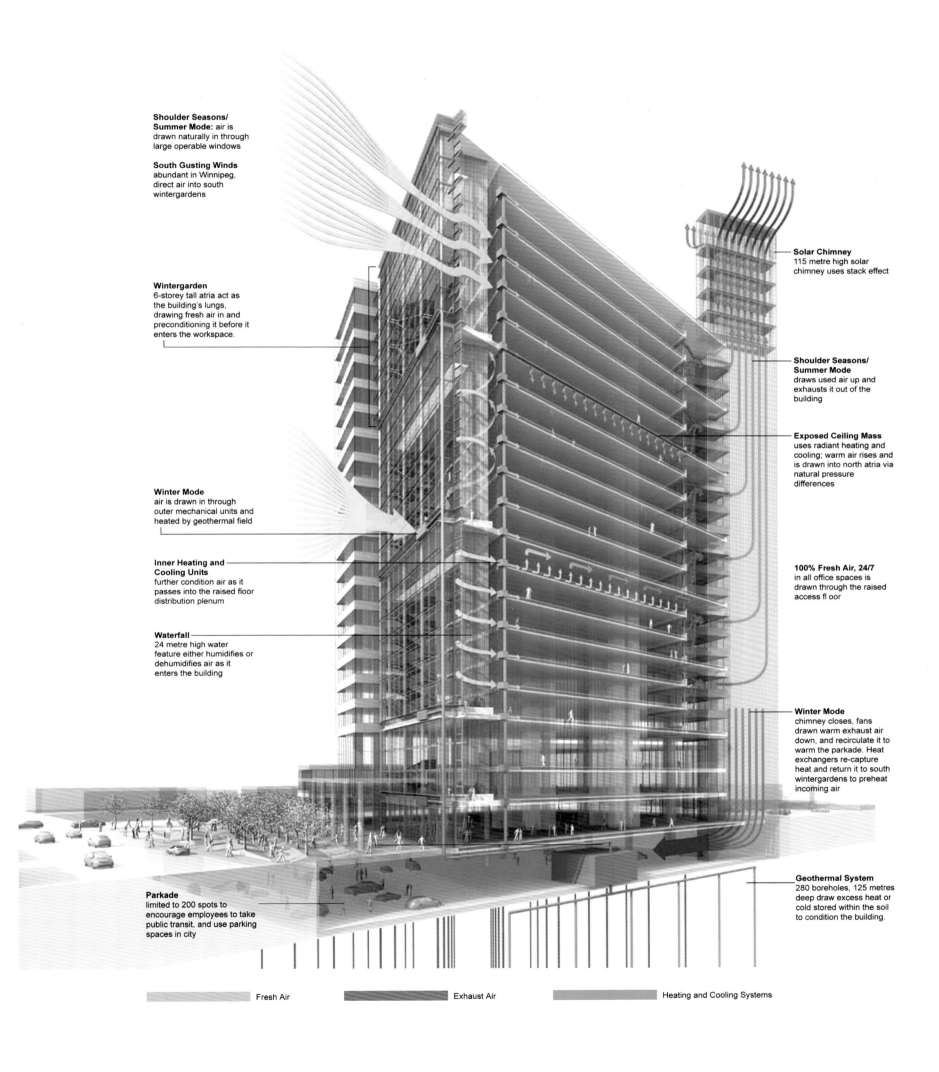

Shape and Structure　　造型与结构

The architectural solution clearly responds to the client's vision, and relies on passive free energy without compromise to design quality and, most importantly, human comfort. The towers converge at the north and splay open to the south for maximum exposure to the abundant sunlight and consistently robust southerly winds unique to Winnipeg's climate. Dubbed an 'Open Book' by citizens of Winnipeg, the tower design forms a capital 'A' composed of two 18-storey twin office towers which rest on a stepped, three-storey, street-scaled podium. In contrast to conventional North American office buildings which use recirculated air, Manitoba Hydro Place is filled with 100% fresh air, 24 hours a day, year round, regardless of outside temperatures. Within the splay of the two towers, a series of three, six-storey south atria, or winter gardens, form the lungs of the building, drawing in outside air and pre-conditioning it before it enters the workspaces through adjustable vents in the raised floor. Depending on the season, a 24 meter tall waterfall feature in each of the atria humidifies or dehumidifies the incoming air. During colder temperatures, recovered heat from exhaust air, and passive solar radiant energy are used to warm the fresh air. A 115 meter (377 feet) tall solar chimney marks the north elevation and main entrance on Portage Avenue, and establishes an iconic presence for Manitoba Hydro on the skyline. The solar chimney is a key element in the passive ventilation system which relies on the natural stack effect. The chimney draws used air out of the building during the shoulder seasons and summer months. In winter, exhaust air is drawn to the bottom of the solar chimney by fans, and heat recovered from this exhaust air is used to warm the parkade and to preheat the incoming cold air in the south atria. Manitoba Hydro Place also has the largest closed loop geothermal system in the province.

该建筑主要通过被动方式来节能，例如材料、朝向等特点，为建筑制冷和采暖提供大部分能源。建筑的两座塔楼形状像大写的字母A，在北部顶端相会，而在南部的底端分开，为的是捕捉充足的阳光。张开的部分容纳了一系列冬季花园，就像肺一样吸纳进新鲜的户外空气，并送到工作场所。每一座中庭都设置了瀑布，根据季节加湿或干燥空气。115 m高的"热烟囱"坐落于主入口，形成了天际线上的标志，也是被动通风的一个关键系统。紧密的热循环系统为办公室采暖和制冷。

KEYWORD 关键词	Skin and Facade 表皮与立面	Glass Facade 玻璃立面
	Materials 材料	Glass 玻璃
	Shape 造型	Spire Model 尖塔造型

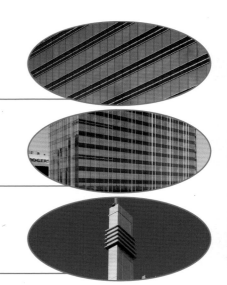

Location: Calgary, Alberta, Canada
Client: Oxford Properties Group
Architectural Design: WZMH Architects
Associate Architect: Gibbs Gage Architects
Landscape Architects: Carson McCulloch Associates Ltd.
Lighting Design: Gabriel Mackinnon

项目地点：加拿大亚伯达卡尔加里
客户：牛津地产集团
建筑设计：WZMH 建筑设计事务所
合作设计：Gibbs Gage 设计事务所
景观设计：Carson McCulloch 设计事务所
灯光设计：Gabriel Mackinnon

Centennial Place
世纪广场

Features 项目亮点

The most important aspect of Centennial Place is its ability to connect with the city on multiple levels. A two storey high pedestrian concourse runs between the buildings, linking to Calgary's unique elevated walkway.

该项目最显著的特点在于其在多个层面连接城市的能力，两层楼高的步行街穿过建筑，将卡尔加里市内独特的高架行人道连接起来。

Overview 项目概况

Centennial Place is the largest commercial LEED Gold development in Canada. The complex's two towers – one of 40 stories and the other 24 stories – provide 97,080 sqm (1,045,000 sqft) of office space.

Perhaps the most important aspect of Centennial Place is its ability to connect with the city on multiple levels. A two storey high pedestrian concourse runs between the buildings, linking to Calgary's unique elevated walkway. WZMH has connected Centennial Place with the street via a row of retail outlets and café/restaurants, creating what will become a bustling new neighborhood when the planned adjacent residential district is completed in the near future. This invigoration of the city's sidewalk culture marks its transformation from commercial to residential neighborhood and a new design paradigm for the public realm in Calgary.

世纪广场是加拿大最大的商业建筑 LEED 金奖开发项目。该项目包括两座塔楼，其中一座 40 层，另一座 24 层，可提供 97 080 m²（1 045 000 平方英尺）的办公空间。

世纪广场最显著的特点在于其在多个层面连接城市的能力。两层楼高的步行街穿过建筑，将卡尔加里市内独特的高架行人道连接起来。通过一条零售商店、咖啡馆及餐馆设计师将世纪广场同街道连接起来，形成一条熙熙攘攘的街道，临近地区为住宅区。令人身心愉快的步行街文化作为商业及住宅区的转换地带，也是卡尔加里市公共区域设施设计的一个新范例。两座塔楼呈直角分布，是对该项目受限制的场地的一次大挑战。

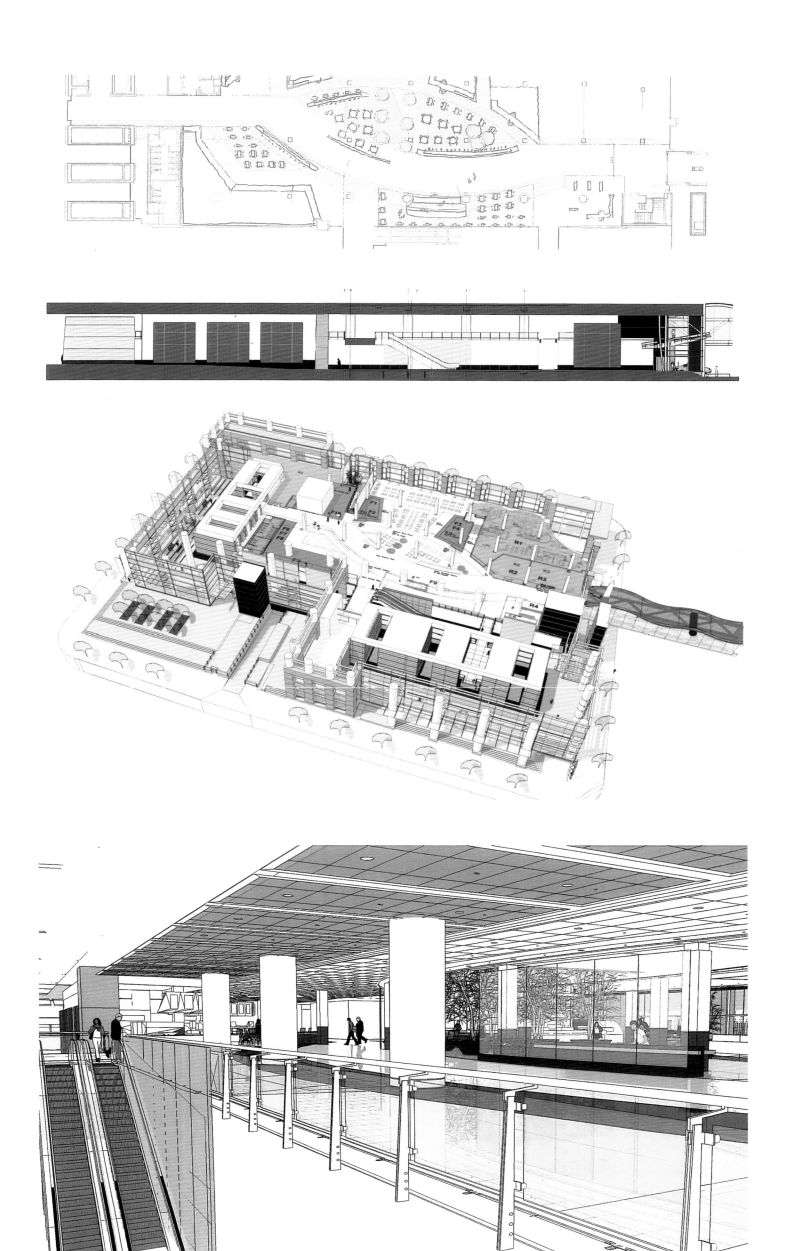

Facade and Skin 立面与表皮

Clad in a high performance glazed façade that minimizes solar ingress and heat gain. Rainwater collection, grey water re-use for irrigation, high efficiency motors in the mechanical plant and occupancy sensors are active elements of a design that also utilizes the building's structural form to create the best internal environments while minimizing energy requirements.

项目的外立面被高性能的玻璃覆盖，最小化了太阳能及热量的吸入。设计外立面被设想成制模过程中的练习，垂直的竖框大小和深度都不同程度地营造出外立面的动态感。

Shape and Structure 造型与结构

The complex's design, described by WZMH as 'dynamic informality', demarcates it from other office towers in the city. The two towers are set at right angles to each other, challenging the regimented convention of this size of development. The external façades of both buildings are envisioned by the architect as an exercise in patternmaking; the vertical mullions being varied in size and depth to create excitement and animation on the immense glazed surface planes of the facades. Additionally, one face of each tower leans inwards as it rises to create 'breathing space' between the two buildings, and, shooting from ground to tip is a spire, which terminates above the highest floor with an illuminated beacon: a focal point from both within and beyond the city limits.

设计中将两座塔都向内倾斜，使两座建筑之间产生了"呼吸空间"。从地面到建筑顶端形成尖塔形状，顶层如同照明灯塔形状，如同市内的天际线。雨水收集系统重复利用污水来浇灌。机械厂和占位传感器的高效电机是该设计中的活跃元素，利用建筑的构造形态来创造最好的室内环境，同时又使能源化需求最小化。该设计被设计师描述成"非正式的动感"，与城市内的其他办公建筑大相径庭。

KEYWORD 关键词		
Skin and Facade 表皮与立面	Glass Facade 玻璃立面	
Materials 材料	Glass, Granite 玻璃、花岗岩	
Shape 造型	Rectangular 矩形外观	

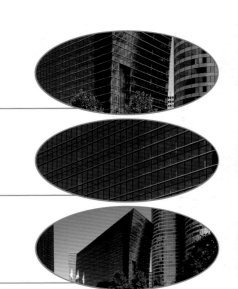

Location: Paris, France
Client: Bouygues Immobilier
Architects: Arquitectonica
Building Area: 23,000 m²
Photo Credit: Paul Maurer

项目地点：法国巴黎
客户：　Bouygues Immobilier
建筑设计：Arquitectonica
建筑面积：23 000 m²
图片来源：保罗·莫勒

Exaltis Office Tower
Exaltis 办公大楼

Features 项目亮点

The building is a rectangular prism modified and sculpted by the introduction of two curves along its short facades. These curving surfaces transform the rigid rectangles into a fluid forms.

大楼整体为长方体造型，两个侧面稍有一些弧度，稍带弯曲的表面让刚硬的矩形外观更富动态感。

Overview 项目概况

A new 15-story 247,570 SF (23,000 m²) office building and 3-level underground parking garage in the La Defense District in Paris. Exaltis is a new office building that completes the axis of the Avenue Gambetta in the La Defense section of Paris. Conceived as a tower of glistening glass, Exaltis defines the axis of the avenue and replaces a grim gray viaduct. The tower is flanked on one side by a landscaped plaza and on the other by a linear park.

Exaltis 办公大楼位于巴黎拉德芳斯区甘贝塔大道，建筑面积为 23 000 m²，包括地上 15 层和地下三层停车库。作为一个闪闪发光的玻璃体，Exaltis 取代了灰色铁架高架桥成为甘贝塔大道的新地标。一个景观广场和一个线性公园分别位于大楼的两侧。

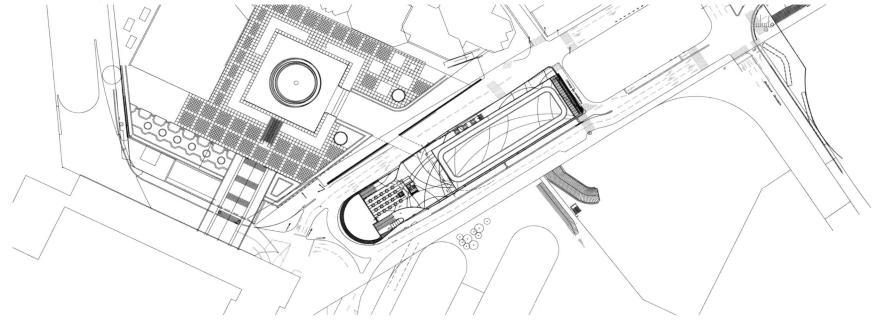

Master Plan

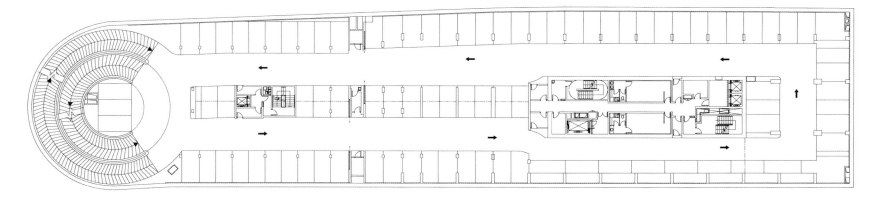

Underground Parking

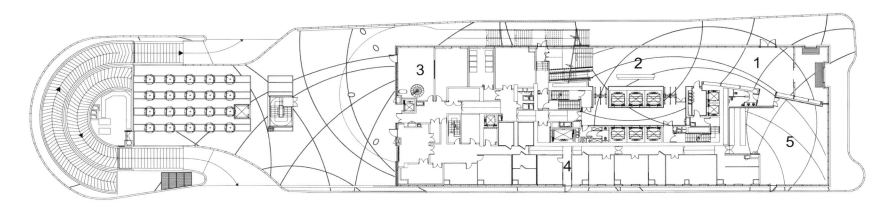

Ground Floor

Legend

1. Lobby
2. Reception
3. Security
4. Kitchen and Pantry
5. Cafeteria

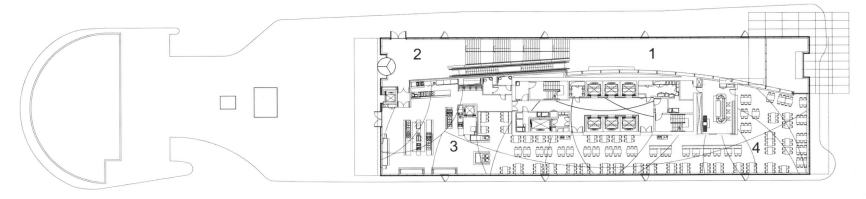

2nd Floor

Legend

1. Open to below
2. Upper Lobby
3. Servery
4. Cafeteria

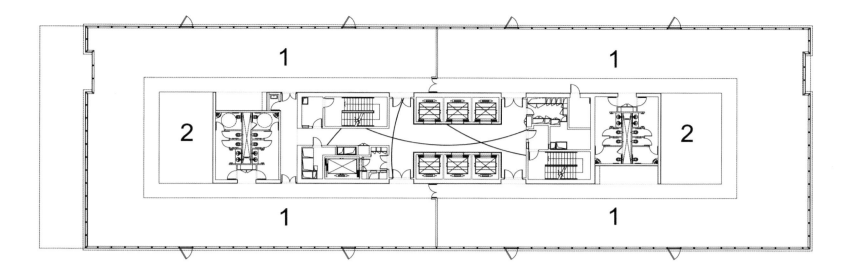

Legend

1. Office
2. Janitor's Closet

Plan - 9th Floor

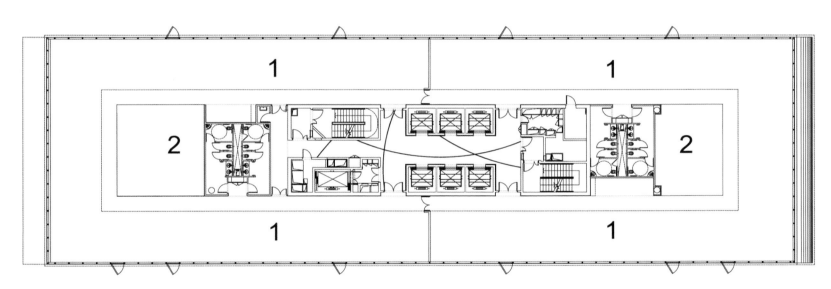

Legend

1. Office
2. Janitor's Closet

Plan - 15th Floor

Shape and Skin 造型与表皮

In form, the building is a rectangular prism modified and sculpted by the introduction of two curves along its short facades. These curving surfaces transform the rigid rectangles into a fluid forms. The two curves radiate from different points somewhere below the ground and splay as they rise towards the sky. The two end walls present contrasting forms, one convex and the other concave. The result is a building with dynamic quality, implying horizontal movement along the axis of the avenue, as if sailing towards the esplanade propelled by an imaginary force.

大楼外观呈长方体，两个侧面稍有一些弧度。弯曲的表面让刚硬矩形外观更具动态型。侧面曲线张开由地面沿伸到天空。两侧面一凸一凹，形成鲜明对比。整个设计动感十足，大楼好似在外力的推动下，如一艘船沿着甘贝塔大道向前航行。

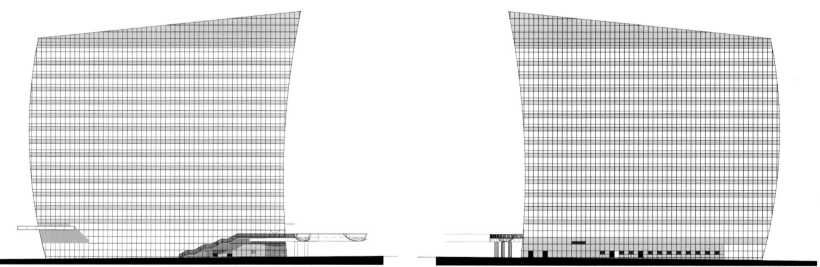

North Elevation   South Elevation

Structures and Materials 结构与材料

Structure: reinforced concrete columns, precast panels; Exterior: curtain wall; wall with white paint, the overlapping curved glass and black granite panels.

结构：钢筋混凝土柱，预制板；外墙：幕墙；墙面采用了白色漆，重叠的弧形玻璃及黑色花岗岩板。

KEYWORD 关键词	Skin and Facade 表皮与立面	Porous Concrete Facade 多孔混凝土立面
	Materials 材料	Concrete 混凝土
	Structure 结构	Diagrid 斜肋构架

Location: Dubai, United Arab Emirates
Architecture Design: Reiser + Umemoto
Site Area: 3,195 m²
Building Area: 27,870 m²

项目地点：阿拉伯联合酋长国迪拜
建筑设计：Reiser + Umemoto
占地面积：3195 m²
建筑面积：27870 m²

O-14
O-14 商业大厦

Features 项目亮点

With porous concrete outer frame structure, the design meets the building's ventilation, environmental protection and energy saving conditions and creates a good visual effect at the same time.

建筑外立面设计采用多孔混凝土外框架结构，形成良好视觉效果的同时满足建筑的通风透气性，环保节能。

Overview 项目概况

O-14, a 22-story tall commercial tower perched on a two-story podium, and comprises over 300,000 square feet of office space for the Dubai Business Bay. O-14 is located along the extension of Dubai Creek, occupying a prominent location on the waterfront esplanade. With O-14, the office tower typology has been turned inside out – structure and skin have flipped to offer a new economy of tectonics and of space.

O-14 商业大厦共22层，建造在一个两层的基座之上。大厦包括约27 870 m²的办公空间，紧临着迪拜湾而建，占据了一个重要的河滨位置，建筑结构和外壁设计都非常有特色。

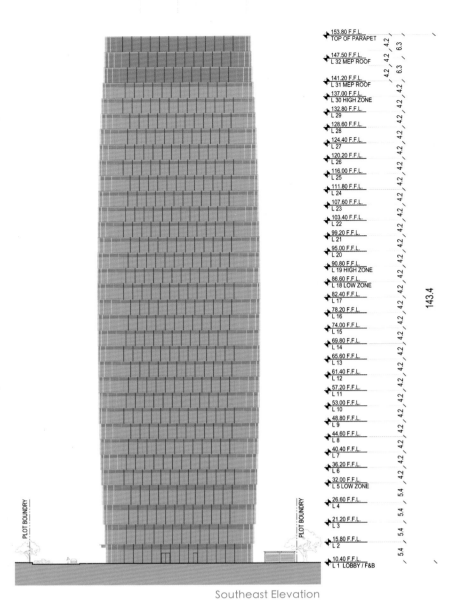

Southeast Elevation

Northeast Elevation

Structure and Materials
结构与材料

Structure: reinforced concrete; Exterior: aluminum frame vitreous.

结构：钢筋混凝土；外墙：铝框架玻璃体。

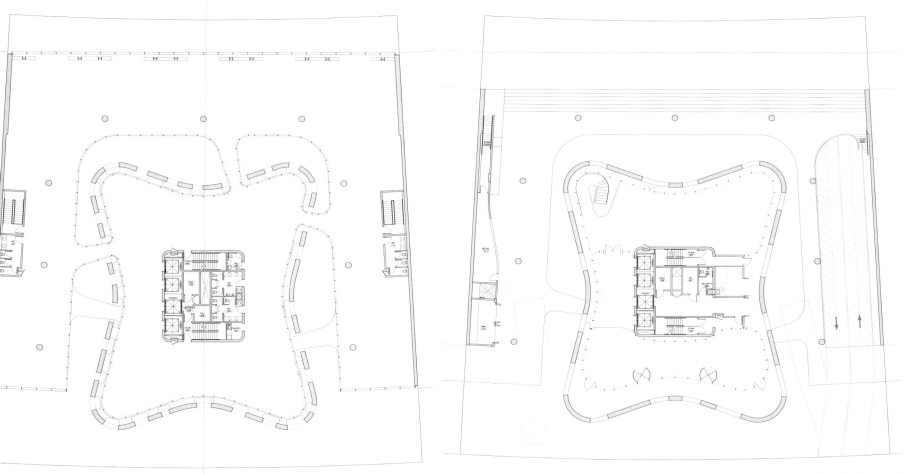

Floor Plan G (Ground)

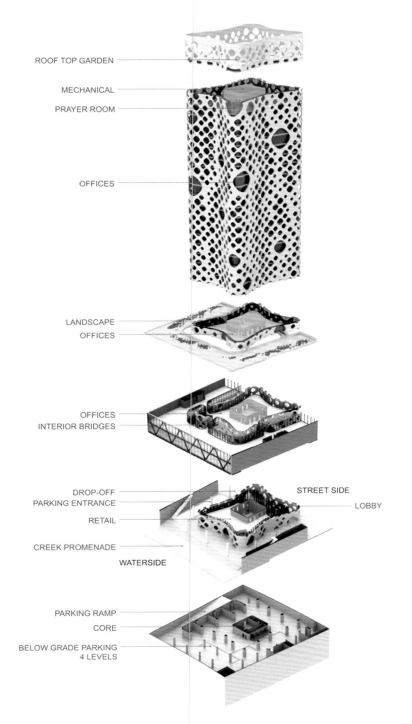

ROOF TOP GARDEN
MECHANICAL
PRAYER ROOM
OFFICES
LANDSCAPE
OFFICES
OFFICES
INTERIOR BRIDGES
DROP-OFF
PARKING ENTRANCE
RETAIL
CREEK PROMENADE
STREET SIDE
LOBBY
WATERSIDE
PARKING RAMP
CORE
BELOW GRADE PARKING
4 LEVELS

TOWER ISOMETRIC

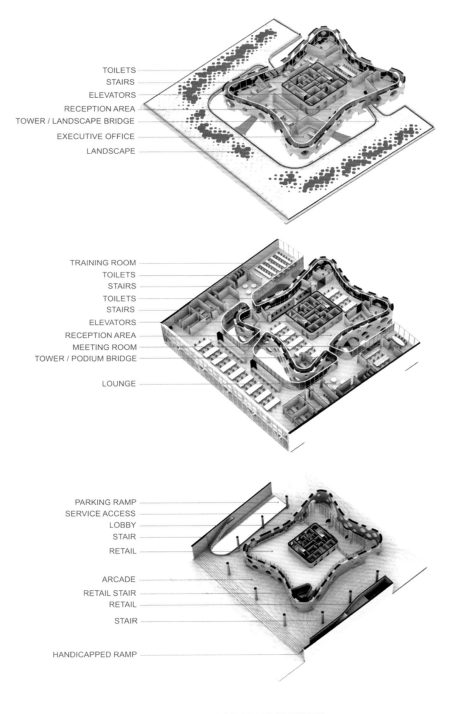

TOILETS
STAIRS
ELEVATORS
RECEPTION AREA
TOWER / LANDSCAPE BRIDGE
EXECUTIVE OFFICE
LANDSCAPE

TRAINING ROOM
TOILETS
STAIRS
TOILETS
STAIRS
ELEVATORS
RECEPTION AREA
MEETING ROOM
TOWER / PODIUM BRIDGE
LOUNGE

PARKING RAMP
SERVICE ACCESS
LOBBY
STAIR
RETAIL
ARCADE
RETAIL STAIR
RETAIL
STAIR
HANDICAPPED RAMP

PODIUM ISOMETRIC

RUR_Isometrics

177

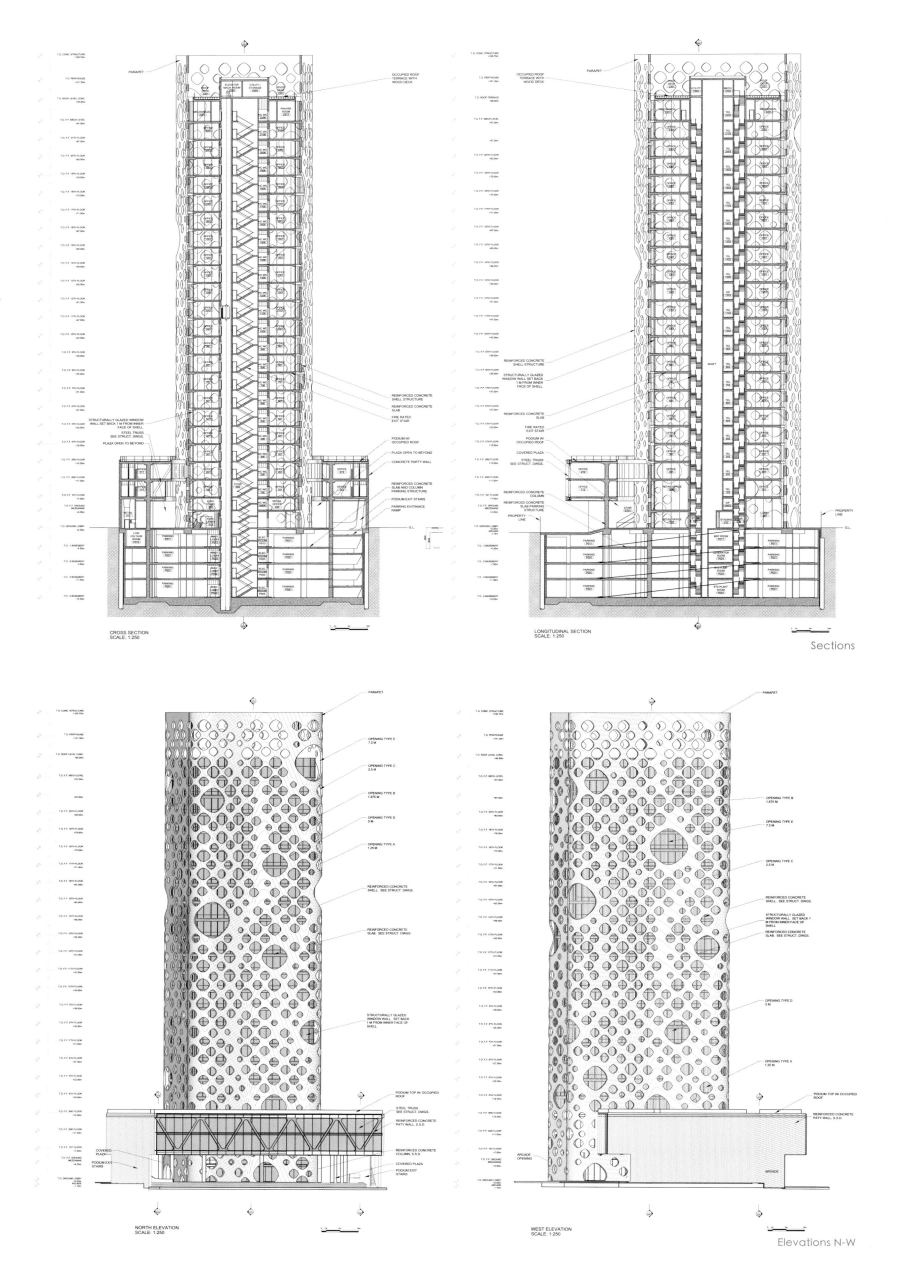

Structure and Skin 结构与表皮

The concrete shell of O-14 provides an efficient structural exoskeleton that frees the core from the burden of lateral forces and creates highly efficient, column-free open spaces in the building's interior. The exoskeleton of O-14 becomes the primary vertical and lateral structure for the building, allowing the column-free office slabs to span between it and the minimal core. By moving the lateral bracing for the building to the perimeter, the core, which is traditionally enlarged to receive lateral loading in most curtain wall office towers, can be minimized for only vertical loading, utilities, and transportation. Additionally, the typical curtain-wall tower configuration results in floor plates that must be thickened to carry lateral loads to the core, yet in O-14 these can be minimized to only respond to span and vibration. Consequently, the future tenants can arrange the flexible floor space according to their individual needs.

The main shell is organized as a diagrid, the efficiency of which is wed to a system of continuous variation of openings, always maintaining a minimum structural member, adding material locally where necessary and taking away where possible. This efficiency and modulation enables the shell to create a wide range of atmospheric and visual effects in the structure without changing the basic structural form, allowing for systematic analysis and construction.

大厦的混凝土外壁提供了一个节能的结构外壳，能够让建筑核心免受侧力影响，并为建筑内创造了高效、无柱的宽敞空间。外壁也是整个建筑最主要的垂直和侧面结构。通过把建筑的侧向支撑移到外部，建筑核心就无需按传统方法扩大来适应幕墙的横向负荷，而是最大程度上缩小来达到垂直负重的要求。另外，标准的幕墙大厦的楼板需要加厚来应付核心的横向负荷，而O-14商业大厦的楼板则无需加厚。这样，大楼空间的使用就可以根据办公的需要来灵活地布局。

建筑外壁也是以斜肋构架进行设计的，它的效能根据开口的连续变化而变化，一般来说它只需要用到少量的结构板。这种外壁设计能够形成良好的透气效果和视觉效果，而不需要改变建筑外观。

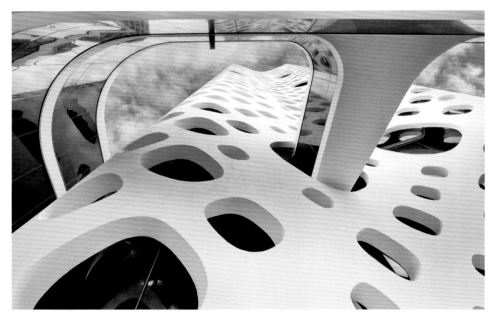

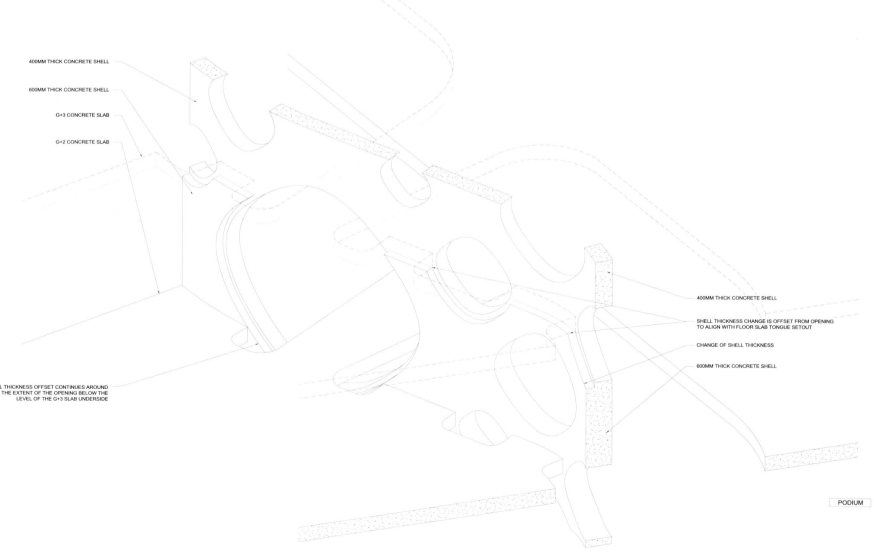

TOWER — Shell Thickness Change Axo

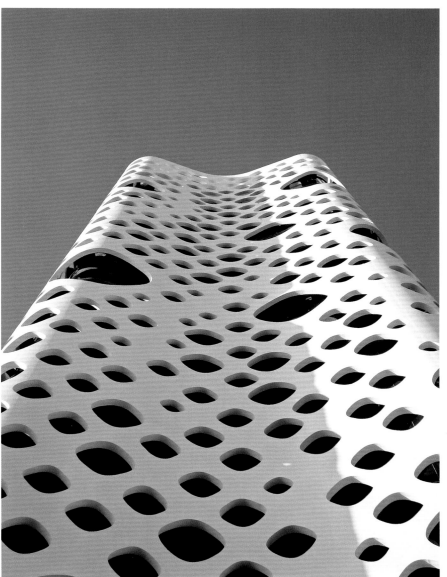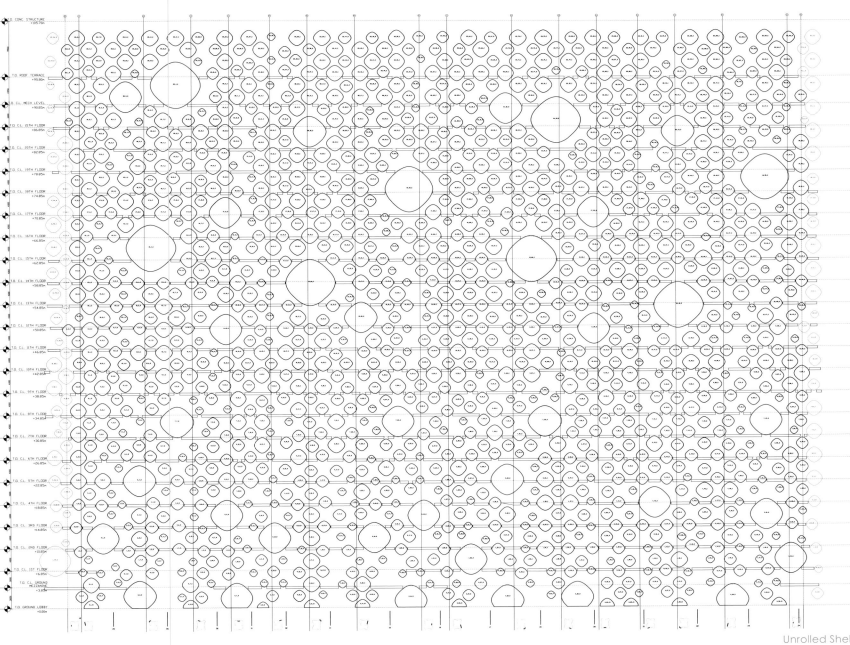

Unrolled Shell-Full

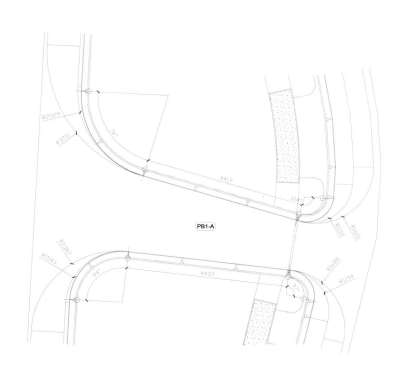
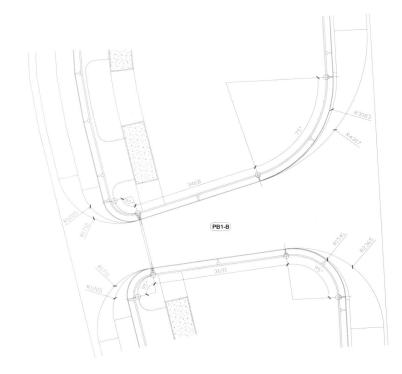

① PB1-A RCP GEOMETRY
A-416 SC 1/50

② PB1-B RCP GEOMETRY
A-416 SC 1/50

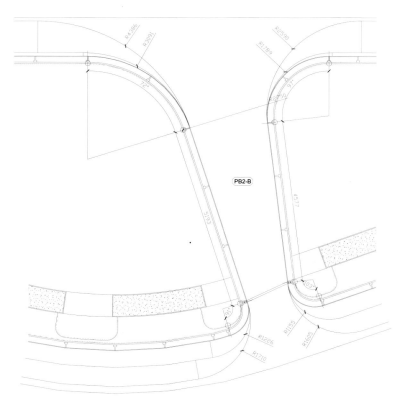
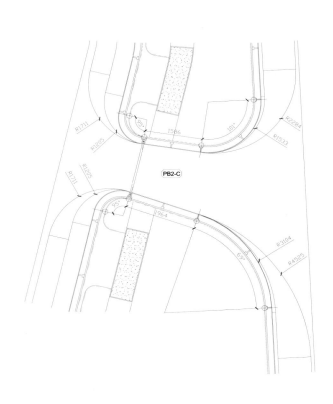

④ PB2-B RCP GEOMETRY
A-416 SC 1/50

④ PB5-C RCP GEOMETRY
A-416 SC 1/50

Detail-Bridge RCP

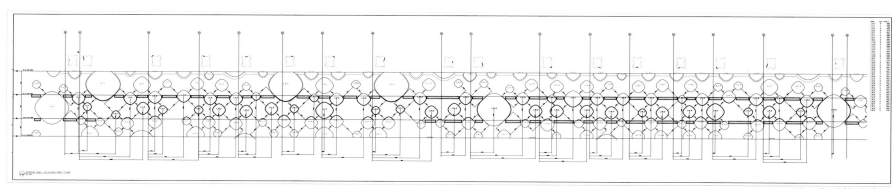

Unrolled Shell G+1

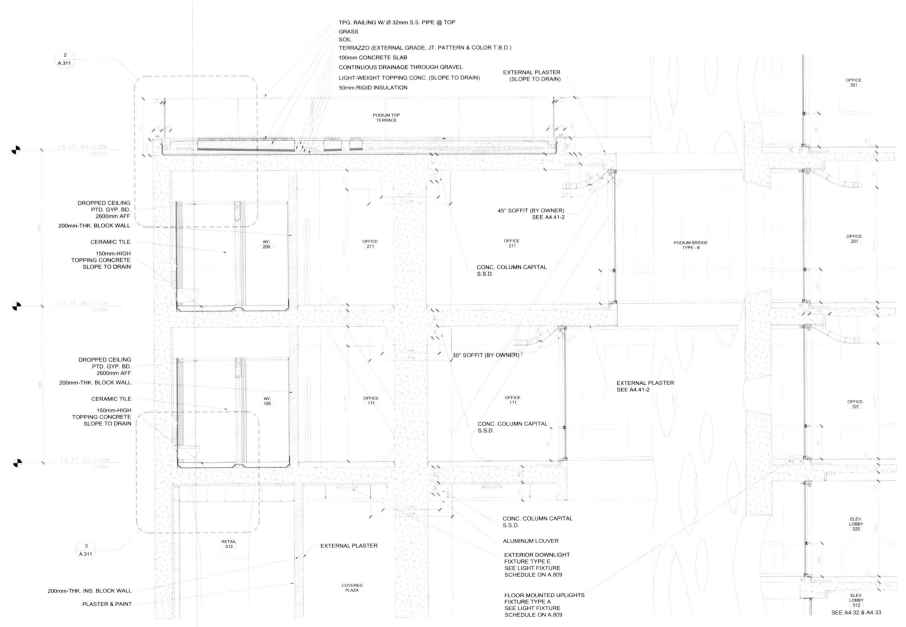

Detail Section - Podium 3

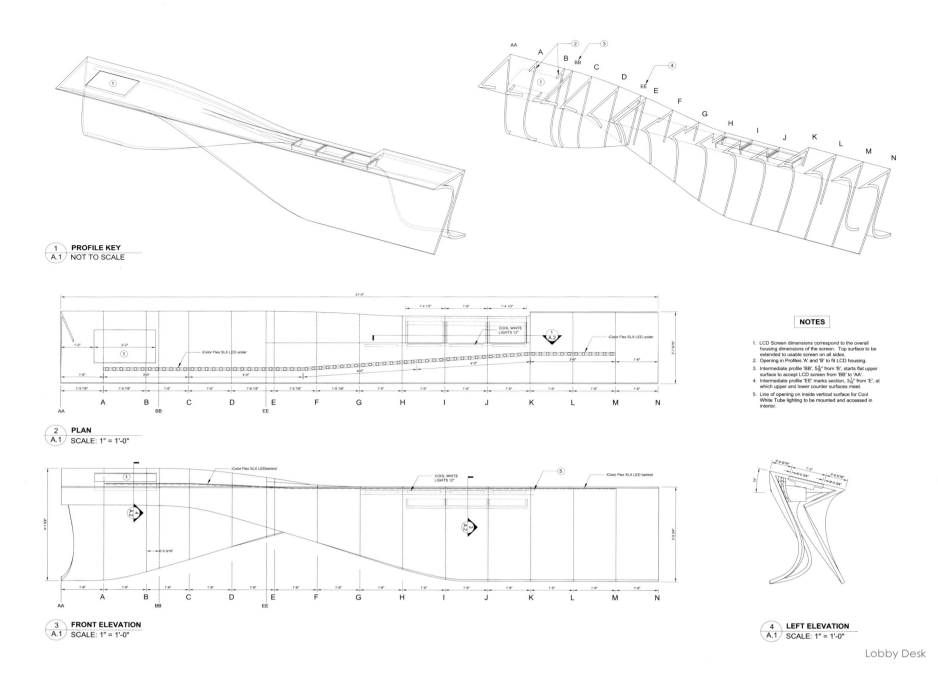

Lobby Desk

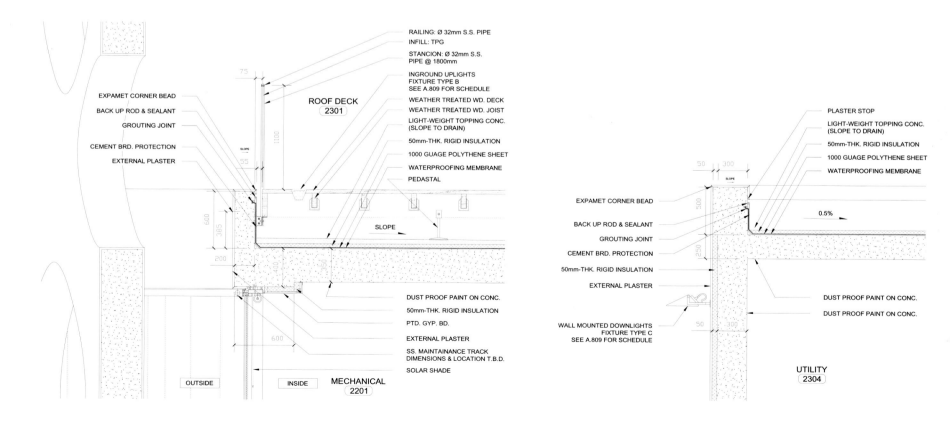

Detail Section - Parapet

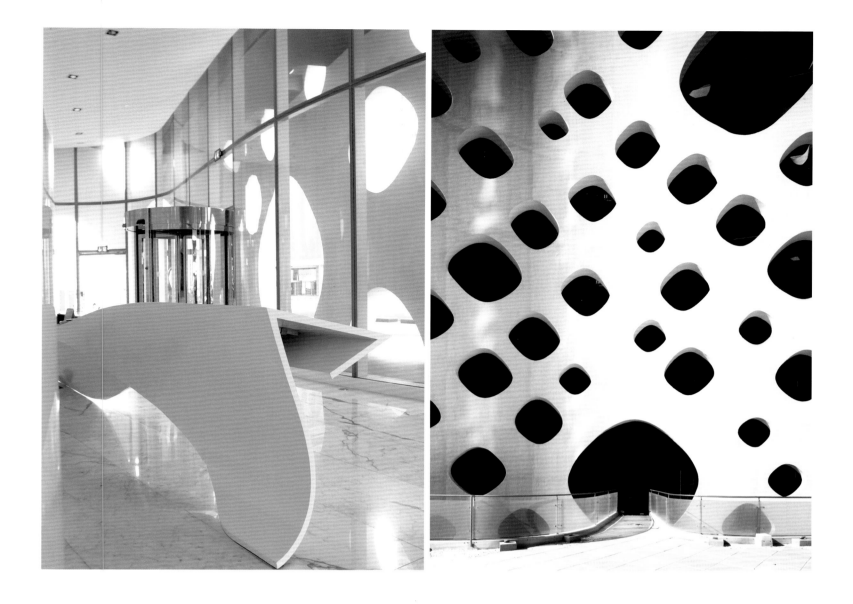
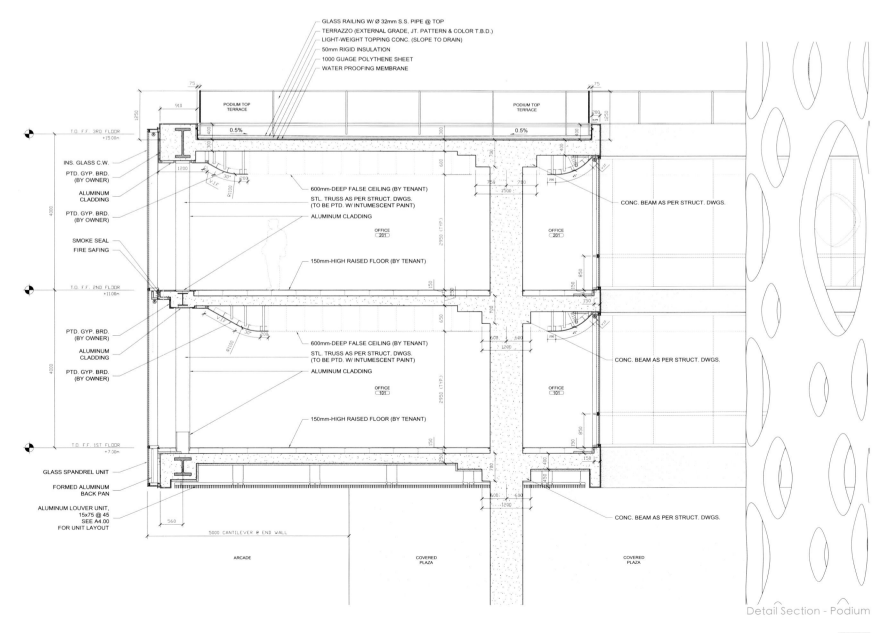

Detail Section - Podium

185

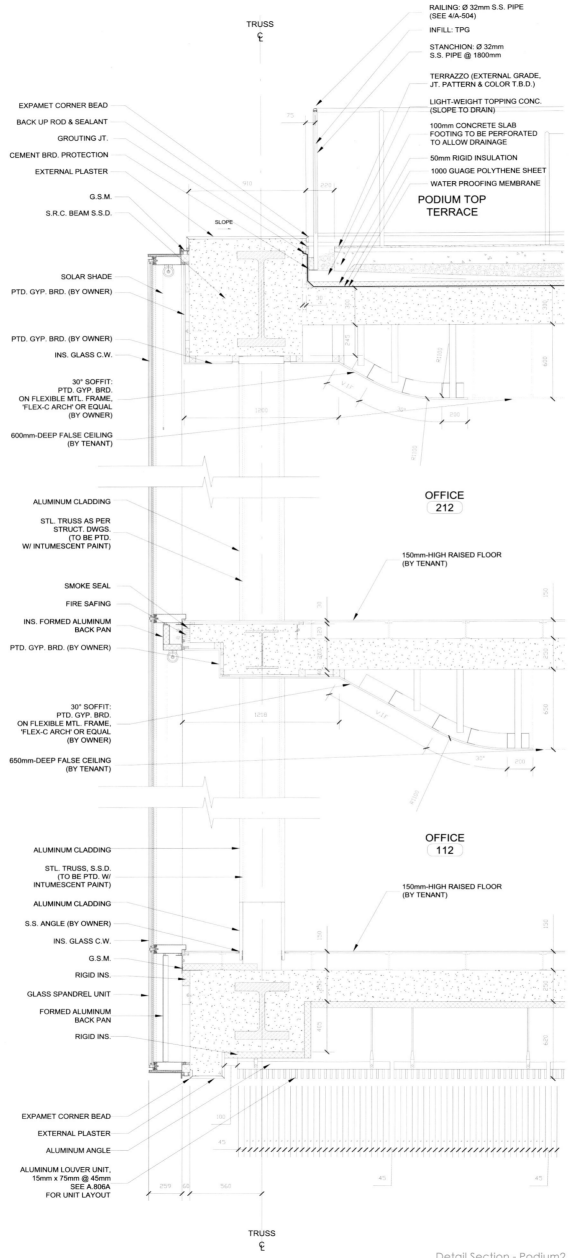

Detail Section - Podium 2

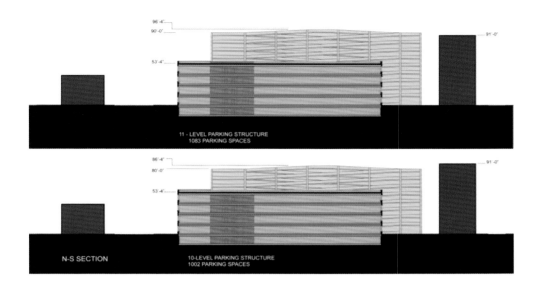

KEYWORD 关键词	Skin and Facade 表皮与立面	Glass Curtain Wall 玻璃幕墙
	Materials 材料	Glass, Steel 玻璃、钢材
	Shape 造型	Fan-Shaped 扇形外观

Location: Makati, the Philippines
Architectural Design: SOM

项目地点：菲律宾马卡蒂
建筑设计：美国 SOM

Zuellig Building
裕利大厦

Features 项目亮点

The fan-shaped floor plates' taper towards the southern face – which allows the building's facades to be set orthogonally to the east and west property lines while also directly engaging the fabric of the city in those directions.

大厦整体建筑外型设计因地制宜，楼体呈扇形，锥尖朝南，这样布局使得楼体外墙呈正东、正西方向，更好地与城市整体结构融为一体。

Overview 项目概况

Rising 33 stories to height of 160 meters, the Zuellig Building is the first premium office tower in the Philippines to be erected since 2000. It is also the first building in Makati that has been pre-certified by the U.S. Green Building Council (USGBC) at the LEED Gold level, and is expected to achieve LEED Platinum certification by September 2013.

Housing nearly 65,000 square meters of Class-A office space and 2,000 square meters of retail space, the highly desirable location at its prominent intersection is intended to accommodate future development. The overall form of the building was derived from the geometry of the site that is bound by Makati Avenue and Paseo de Roxas.

共33层，高160 m的裕利大厦，是菲律宾2000年以来新建的第一个优质写字楼。它也是马卡蒂第一个通过美国绿色建筑委员会（USGBC）的LEED金级预认证的大楼，并有望在2013年9月获得LEED铂金认证。

裕利大厦拥有A级办公面积近65 000 m²，零售面积2 000 m²，坐落于Makati大道和Paseo de Roxas街的交汇点，地理位置优越，发展前景广阔。

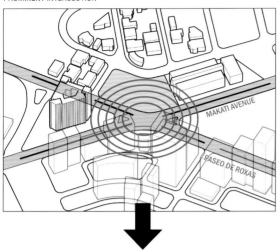
PROMINENT INTERSECTION

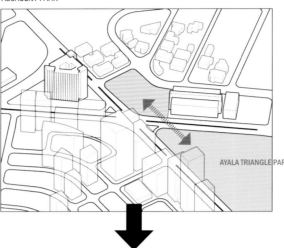
ADJACENT PARK

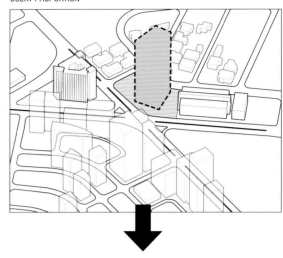
BULKY PROPORTION

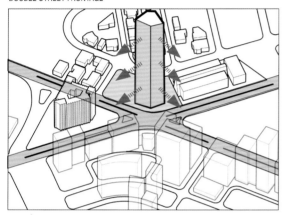
DOUBLE STREET FRONTAGE

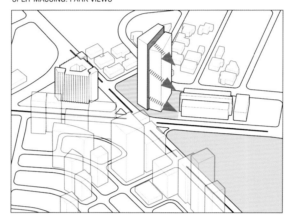
SPLIT MASSING: PARK VIEWS

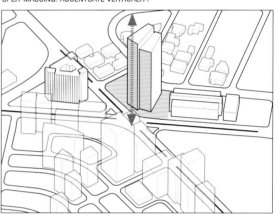
SPLIT MASSING: ACCENTUATE VERTICALITY

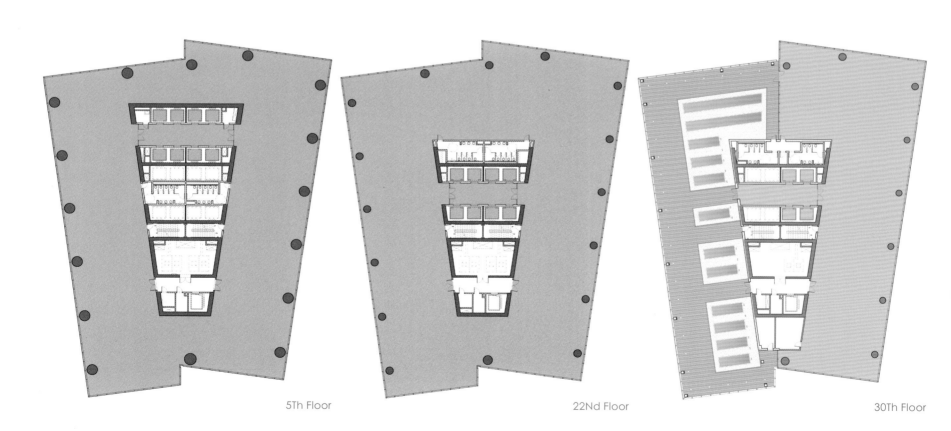

5Th Floor   22Nd Floor   30Th Floor

Style and Structure 造型与结构

The fan-shaped floor plates' taper towards the southern face – which allows the building's facades to be set orthogonally to the east and west property lines while also directly engaging the fabric of the city in those directions. Each floor plate offers between 1,870 and 1,945 square meters of leasable office space per floor with a column-free layout that provides maximum efficiency and flexibility. A raised floor system at all levels allows modular office layout modifications and conduits to facilitate upgrades in computer and communications technologies. A double-height open-air terrace on the 30th floor provides outdoor gathering space for tenants and visitors. A low-rise pavilion attached to the tower houses two street-level retail lots, a food court, and an executive lounge containing a restaurant and meeting rooms. Open air terraces provide views to the Ayala Triangle Park and Urdaneta Village.

大厦整体建筑外型设计因地制宜，楼体呈扇形，锥尖朝南，这样布局使得楼体外墙呈正东、正西方向，更好地与城市整体结构融为一体。各楼层提供 1870~1945 m² 不等的办公出租空间。无柱的空间结构使得空间利用率和灵活性最大化。所有楼层都配置有活动地板系统，极大地便利了办公布局、管道调整及计算机和通信技术的升级改造。位于30层的加高露天阳台为租户及访客提供户外聚会空间。大楼底层有两个零售亭、一个食品店、一个含餐厅和会议室的行政酒廊。露天露台可欣赏阿亚拉三角公园和乌达内塔村的风景。

KEYWORD 关键词	Skin and Facade 表皮与立面	Glass Façade 玻璃立面
	Materials 材料	Glass, Metal Framework 玻璃、金属框架
		Detail 细部

Location: Osaka, Japan
Client: Fukoku Mutual Life Insurance Company, Osaka, Japan
Architect: Dominique Perrault Architecture, Paris, France
Site Area: 3,900 m²
Built Area: 68,500 m² (including car park)
Maximum Height: 133 m
Photography: Daici Ano / DPA / Adagp
　　　　　　 Dominique Perrault / DPA / Adagp
　　　　　　 DPA / Adagp

项目地点：日本大阪
客　　户：日本大阪富国相互生命保险公司
建筑设计：法国巴黎 Dominique Perrault 建筑师事务所
基地面积：3 900 m²
建成面积：68 500 m²（包括停车场）
最大高度：133 m
摄　　影：Daici Ano / DPA / Adagp
　　　　　Dominique Perrault / DPA / Adagp
　　　　　DPA / Adagp

Fukoku Tower
大阪富国大厦

Features 项目亮点

Taking inspiration from the profile of a gigantic tree whose roots proliferate on the surface of the ground, splayed at its base, the tower's outline tapers elegantly as it rises, gracing the city's skyline with a vertical asymptote.

设计灵感源于一颗巨树，基座部分的镶镜立面为扩散状，与上部光滑的玻璃立面形成鲜明对比，极具标识性。

Overview 项目概况

This tower project for the Fukoku insurance company takes inspiration from the profile of a gigantic tree whose roots proliferate on the surface of the ground. Splayed at its base, the tower's outline tapers elegantly as it rises, gracing the city's skyline with a vertical asymptote. There are 32 floors in total: 28 floors on surface, 4 basement levels. Office space is 30,000 m², university laboratories space is 3,400 m², office lobby is about 950 m², car park is about 4,080 m² and commercial area is 4,930 m².

在日本发达都市大阪的中心，像水晶一样的富国相互生命保险公司大楼冲天而起。该大厦的设计灵感来自一棵巨大的树，其根部在地面扩散。大厦共32层，地上28层，地下4层，其中办公面积30 000 m²，大学实验室面积3 400 m²，办公楼大堂面积950 m²，停车场面积4 080 m²，商业区面积4 930 m²。

Hotel Building
酒店建筑

Nowadays, hotel buildings are not only used for accommodation and dinning but also serve the business, convention, holiday, recreation and so on. According to their different functions, hotel buildings can be classified into business hotels, holiday hotels, apartment hotels, convention hotels, resort hotels, etc. With the application of new technologies and new materials, the styles of hotel buildings have changed innovatively from shape, facade as well as interior decorations.

The building shape is decided by the cultural background and management philosophy. The design should clearly highlight the culture and style of the hotel brand. Reinforced concrete structure is commonly used for hotel buildings. While the facade is usually designed with novel glass, metal plate, precast concrete panel, etc. The interior decoration combining with exterior exterior design shapes the integrated environment for the hotel. Thus the interior design of the hotel should keep harmonious with its exteriors.

今日的酒店建筑，功能已从提供单纯的住宿和餐饮服务，发展至商务、会议、度假、休闲服务等众多方面。根据不同的性质，酒店可分为商务型酒店、度假型酒店、公寓式酒店、会议型酒店、观光型酒店等等。随着新技术、新材料的发展应用，以及酒店类型、风格的不断推陈出新，酒店建筑设计无论从建筑造型、立面到室内装饰设计都有了巨大的变化。

酒店建筑因酒店本身不同的文化主题，不同的经营理念，在其造型设计上有很大的不同，设计时应清晰准确地体现出酒店的文化内涵和风格。酒店建筑的结构以常见的钢筋混凝土结构为主。立面常采用玻璃立面、金属立面、石材立面等，材料则根据不同的设计需求，选择各类新型玻璃、金属材质面板、预制混凝土面板等。酒店建筑的室内装饰设计与酒店外部的设计共同组成整个酒店的环境，因而，在进行室内装饰设计时应充分考虑酒店的外部设计，以达到酒店内外环境的和谐统一。

KEYWORD 关键词

- Skin and Facade 表皮与立面 | **Metal Facade** 金属立面
- Materials 材料 | **Aluminum Panel** 铝制面板
- **Cantilever Structure** 悬臂结构

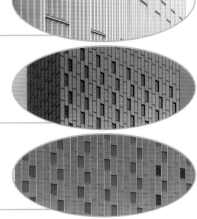

Location: Barcelona, Spain
Client: Hoteles Sol-Melia
Architects: Dominique Perrault Architecture
Partner Architects: Corada Figueras Arquitectos
Local Architect: AIA Salazar-Navarro
Total Floor Area: 29,334 m²
Photography: André Morin, Miguel de Guzman

项目地点： 西班牙巴塞罗那
客户： Hoteles Sol-Melia
建筑设计： 法国 Dominique Perrault 建筑师事务所
合作设计： Corada Figueras Arquitectos
本地建筑师： AIA Salazar-Navarro
总建筑面积： 29 334 m²
摄影： Andre Morin, Miguel de Guzman

ME Barcelona Hotel
西班牙巴塞罗那 ME 酒店

Features 项目亮点

The tower is composed of two volumes stuck together. The way these boxes are placed against each other creates a brand new city landmark.

大楼由两个盒子状体量"粘"在一起，这些体量互相叠放的设计，创建出一处崭新的城市地标。

Overview 项目概况

Designed for the Habitat group in Barcelona and now managed by ME, this hotel integrates the two dimensions that compose the identity of the Catalonian capital: the horizontality of its grid, legacy of the Cerda plan, extending all the way to the sea, and its dynamic verticality exemplified by the Sagrada Familia and Mount Tibidabo looming over the sight.

The hotel is compound by 259 guest rooms(192 supreme rooms, 44 superior rooms, 16 suites, 6 grand suites, 1 sky suite, 4 double rooms with access for disabled people), a fitness, a restaurant (300 m²), a conference centre (1,150 m²), a swimming pool and terraces, a bar, salons, administration and underground car park.

这家酒店原先为加泰罗尼亚首府巴塞罗那的 Habitat 集团设计，如今归 ME 接管。酒店集构成巴塞罗那城市特色的两大方面于一身：一是城市的水平网格，这是塞尔达城市规划的遗迹，所有道路都通向大海；另一个就是动态的垂直风景，笼罩视野的圣家族大教堂和蒂维达沃山就是最佳实例。

酒店由 259 间客房（192 间尊贵套房、44 间高级客房、16 间套房、6 间大套房、1 间天空套房，其中 4 间双人房设有残障人士通道）、健身室、餐厅（300 m²）、会议中心（1 150 m²）、游泳池和露台、酒吧、多个沙龙、行政管理区和地下停车场组成。

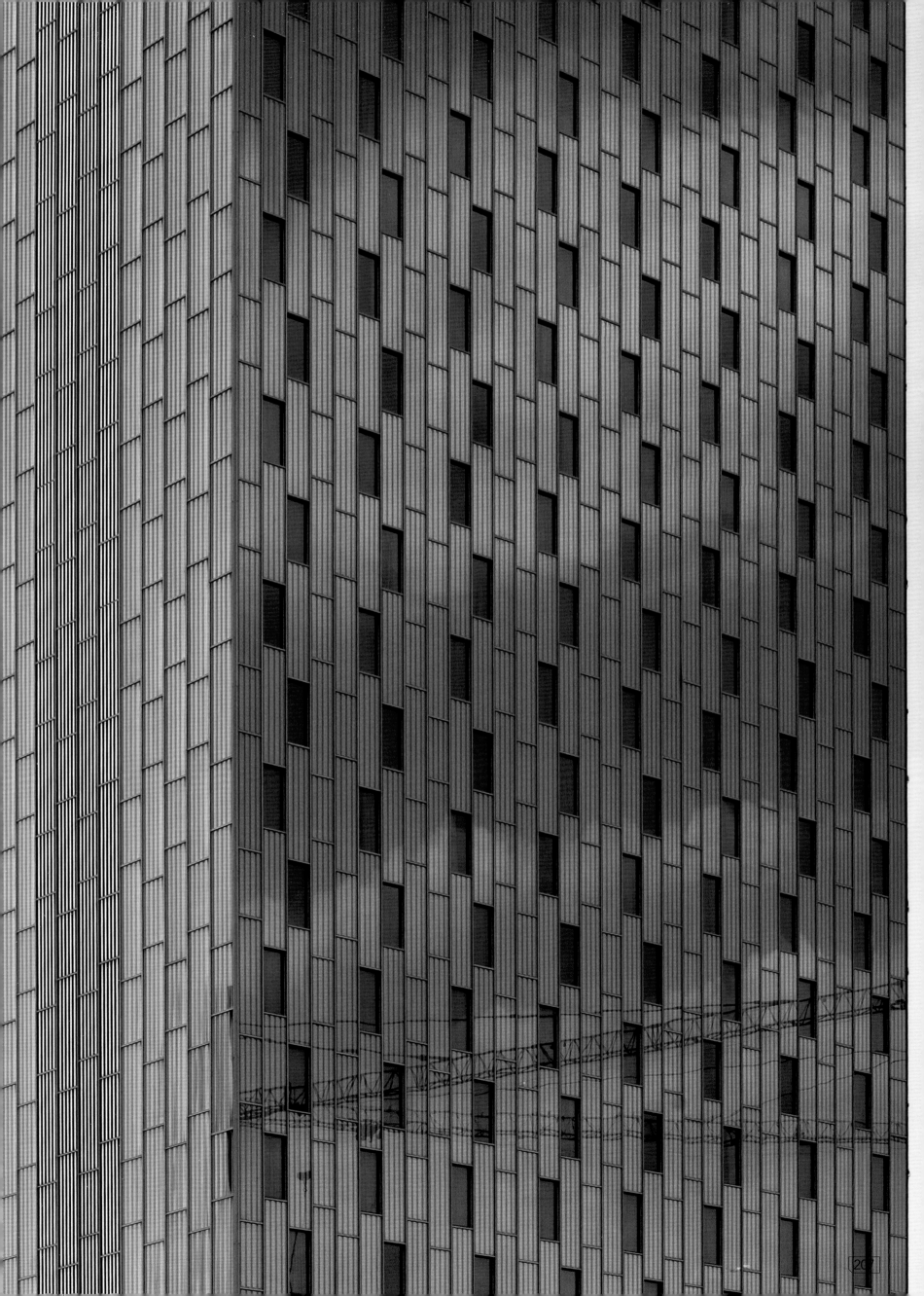

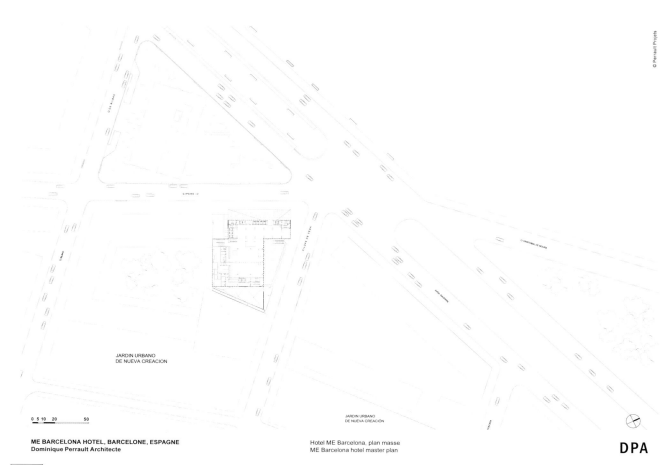

ME BARCELONA HOTEL, BARCELONE, ESPAGNE
Dominique Perrault Architecte

Hotel ME Barcelona, plan masse
ME Barcelona hotel master plan

DPA

Facade and Structure

The tower is composed of two volumes stuck together: a "cubic" building acting as a counterpoint and a tower 120 m high, a rectangular parallelepiped cut lengthwise in two. A cantilever, 20m above street level, marks the entrance: on the Avinguda Diagonal it serves as the Hotel's identifying signal.

The way these boxes are placed against each other is key to the distribution of the various functions. While the volume located at the back gathers the hotel's collective services, the tower, broad but not deep, houses the 259 guest rooms, each with a clear perspective of the scenery. This "enormous screen that focus on the city and the landscape" is cut into opaque panels of distinct texture that cover the entire facade, making it come alive in the day and the night.

An arrangement of elementary forms creates the building's reference signs: a 25 meter high canopy in the style of a loggia points to the Tower; a protuberance in a form of a cantilever creates a "crest" in the vertical city skyline; the cube shifts back to free up a small square like a terrace that opens onto Calle Lope de Vega. This creates a new point of reference in the more recent part of the Diagonal, with the tower standing out against the sky.

The combination of these urban signs gives the Tower a real capacity for architectural interaction with the present and future context of the area. The functional organisation is the logical consequence of its architectonical situation. At the base of the building are the activities linked to movement and meeting such as the hotel lobby, the restaurants, the meeting rooms, the swimming pool and the day and night bars. The main unit holds the individual and double rooms, and the suites that open onto the sea or the mountain, with views towards the Sagrada Familia. The interior design and comfort of the hotel are based on the generous views from each room, like a giant screen overlooking the city landscape. This screen is articulated by a series of smaller screens in the manner of television sets, which form a "wall of images". The result is a building clad in an armour of aluminium sheets.

The tower will stand out in the Barcelona skyline like a metal needle; a lively, happy "jewel", with red, blue and green glass distributed at random along the facade like a giant stained glass window. At night, the tower turns into an "urban la tern, a luminous symbol of the Diagonal.

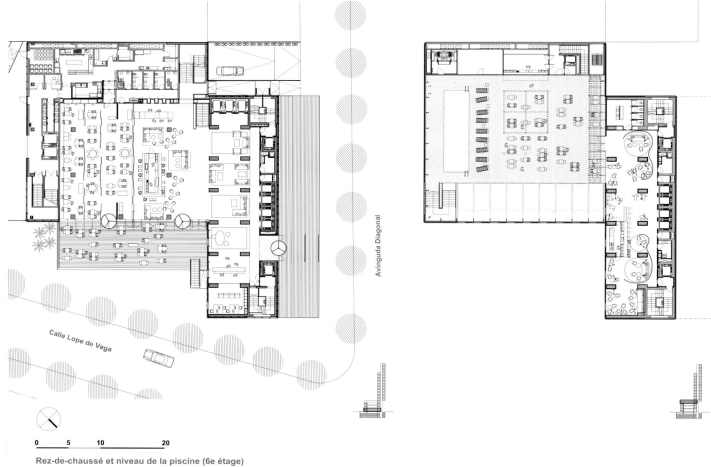

Rez-de-chaussé et niveau de la piscine (6e étage)

Ground floor and swimming pool (6th floor)

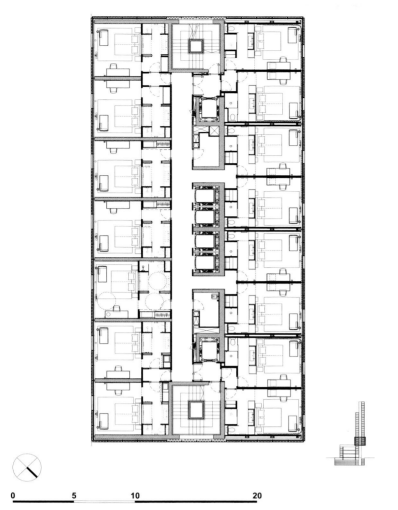
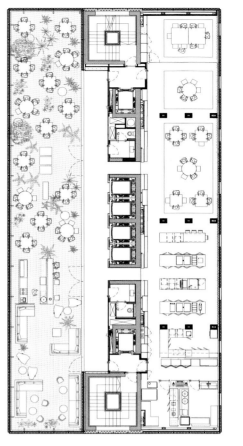

Etages des chambres (7e au 16e étage) et étage du Sky restaurant (24e étage)
Hotel room floor (7th to 16th floor) and Sky restaurant floor plan (24th floor)

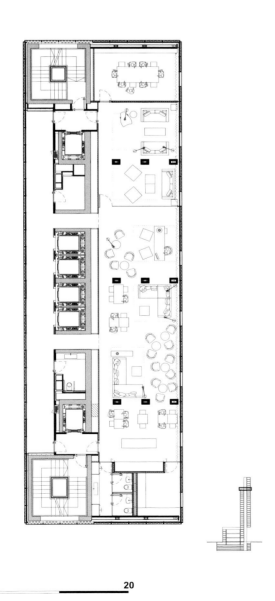

Etages Executive (25e étage) et étage des suites (26e au 28e étage)
Executive floor (25th floor) and suites floor plan (26th to 28th floor)

211

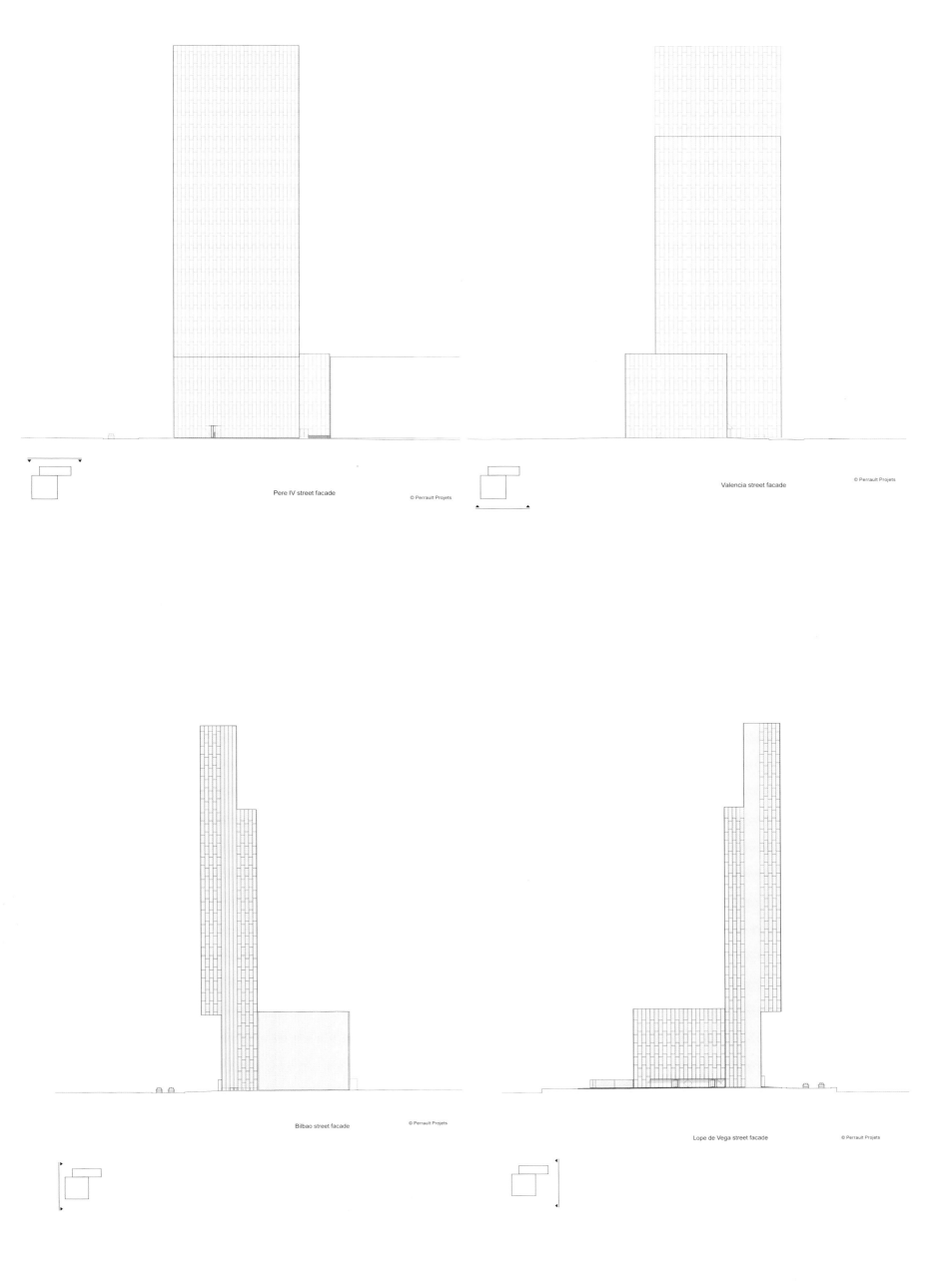

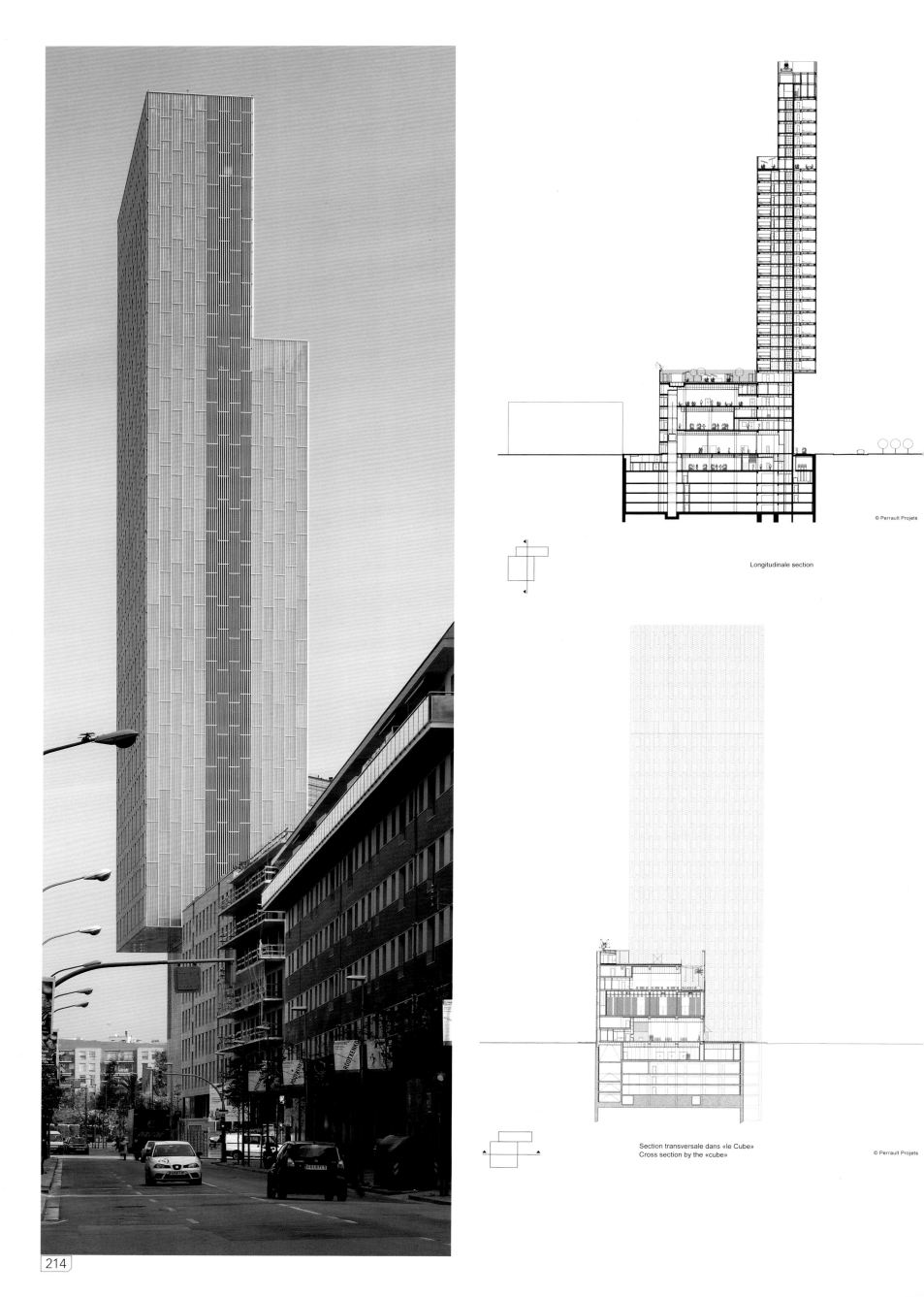

Longitudinale section

Section transversale dans «le Cube»
Cross section by the «cube»

立面与结构

这座大楼由两个体量"粘"在一起：一座立方体建筑作为基座，一座120 m高的长方体塔楼在纵向的两侧分别切去一块。一个距地面20m高的悬臂结构就是入口，在Avinguda Diagonal大道上，让人一眼就能认出酒店所在的位置。

这些盒子状体量互相叠放的方式是分布各种功能的关键。虽然位于后方的体量将酒店的各种服务功能集中在一起，但是进深虽不大却宽敞的塔楼容纳了259间客房，从每个房间都能清晰地欣赏到窗外美景。这种"巨大的立面设计将重点放在城市与景观欣赏上"，被切成纹理截然不同的不透明面板，覆盖着整个立面，无论白天黑夜，酒店立面看起来都活力四射。

基本形式的排列形成了建筑的标志性特征：塔楼顶端25 m高的顶棚被设计为凉廊的形式；突出的悬臂结构在垂直的天际线上创造出一个"羽冠"；立方体建筑向后留出一块如同露台的小广场，面向Calle Lope de Vega大道开放。塔楼向着天空拔地而起，如此设计将在Diagonal大道创建一处崭新的地标。

这些城市标志的组合使塔楼能够在当前及未来的区域文脉中，真正在设计上具有与其他建筑相互呼应的能力。功能结构是根据建筑环境而进行逻辑推断设计的成果。在建筑基础部分设有酒店大堂、餐厅、会议室、游泳池和昼夜酒吧等运动与聚会场所。主要结构容纳各种单人间、双人间和套房，依山面海，可俯瞰圣家族大教堂。酒店的室内设计与舒适度建立在每间房间均拥有广阔视野的基础上，就如同一面巨大的屏幕，俯瞰着城市风光。这面屏幕由一系列小屏幕以电视机的方式连接而成，形成了一面"图片墙"。因此形成了一座身披铝板盔甲的建筑。

这座塔楼在巴塞罗那的天际线如一根金针一般脱颖而出；也似一颗富有朝气、令人愉悦的"宝石"，用红、蓝、绿色玻璃随机分布在立面上，就像一扇巨大的彩色玻璃窗。到了晚上，塔楼就变成了一座"城市灯塔"，是Diagonal大道的璀璨象征。

KEYWORD 关键词	Skin and Facade 表皮与立面	Glass Facade 玻璃立面
	Materials 材料	Serigraphed Glass 绢网印花玻璃
		Steel-Frame Arm 钢架悬臂

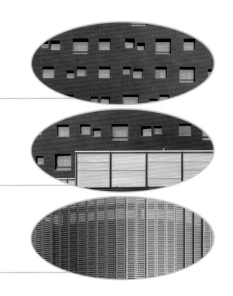

Location: Milan, Italy
Architect: Dominique Perrault Architecture, Paris
Site Area: 15,000 m²
Built Area: 23,800 m²
Photography: © André Morin / DPA / Adagp

项目地点：意大利米兰
建筑设计：法国巴黎 Dominique Perrault 建筑师事务所
占地面积：15 000 m²
建成面积：23 800 m²
摄　　影：© André Morin / DPA / Adagp

NH-Fieramilano Hotel
米兰国际展览中心 NH 酒店

Features 项目亮点

Each tower is inclined at an angle of 5 degrees and being connected by two large semi-transparent arms, form a simple yet powerful landmark.

两座酒店均有 5 度倾斜角，以半透明的悬臂结构连接，形成简单却强势的地标形象。

Overview 项目概况

The project for hotel in the new Trade Fair in Rho-Pero, Milan, fits in with the city's great architectural-building tradition. Two pure geometric forms, which are sixty meters tall and set together at an angle, emerge from the horizontal landscape of the new trade fair. They form a simple yet powerful landmark. The radical pureness of the two buildings embodies the most deeply entrenched and profound of Milanese culture; sobriety and simplicity.

该项目包括两个纯几何形状的酒店，高约 60 m，位于新的贸易展览场地内，与米兰杰出的建筑传统相得益彰，形成了简单却强势的地标景象，也体现出米兰文化的特质：冷静与朴素。

Structure And Materials 结构与材料

The canopy is made of steel portals supporting a light roof made of serigraphed glass.

悬臂顶棚由钢架构成，支撑着绢网印花玻璃制成的轻质屋面。

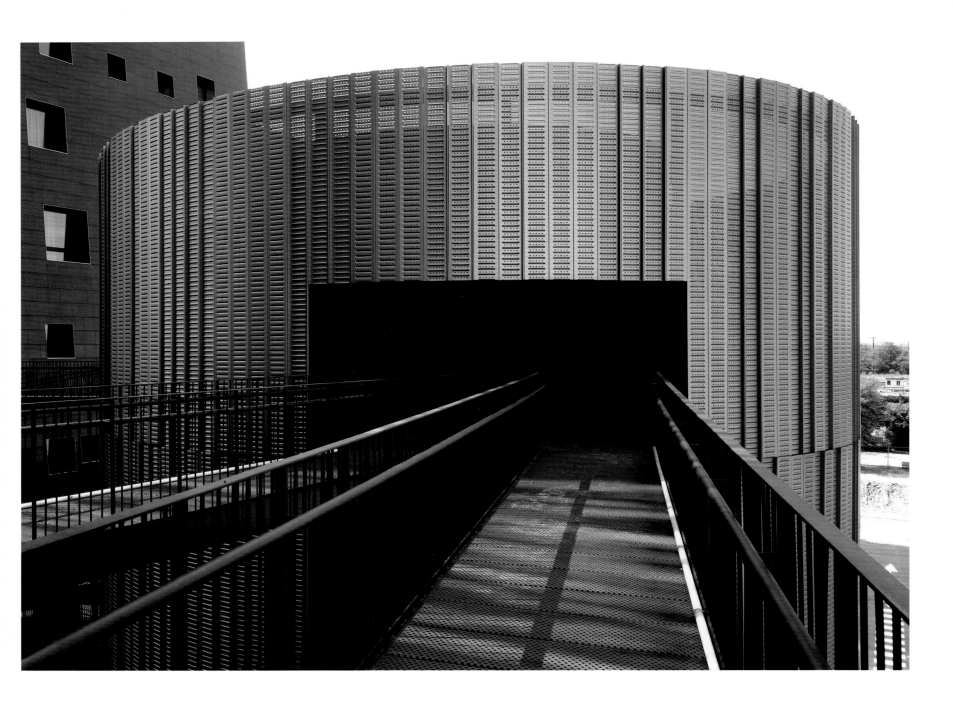

Facade and Shape 立面与造型

The two structures are located at the main hub of the orthogonal plan. Each tower is inclined at an angle of 5 degrees: the taller tower facing the Trade Fair to the north, the second leaning over mark the pedestrian entrance way. The cross is formed by two large semi-transparent arms. The transparency of the walls varies with the light and always flows in one direction. The cross sets out the entrances to the floors of the two towers holding the rooms and also the corridors connecting the various refreshment services and facilities: bar, brasserie, restaurant, banquet rooms and meeting rooms are easy to identity and reach through the spacious lobby, which extends seamlessly into the towers through two wide gaps. The main entrances to the reception facility are at two ends of the arms of the cross. The reception is near the centre and instantly visible and recognisable. The bases of the tower are at the two other ends, with two entrance ramps to the levels holding the communal facilities and, above them, the floors of 4-star rooms.

两座酒店位于正交平面的主要部位，每座都有5度的倾斜角：稍高的一座背向展馆，另一座标志着行人入口。连接两座酒店的十字架是半透明的悬臂，其透明度随灯光的变化而变化（单向的），方便酒店客户往来其间，共享部分设施：酒吧、餐厅，宴会厅和会议室。通往接待区的主入口位于十字架其中两个末端，接待区靠近展览中心，很容易识别与辨认。酒店的底部位于另外两个末端，此处有入口匝道引至公共设施楼层，往上便是四星级酒店空间。

KEYWORD 关键词

Skin and Facade 表皮与立面 | **Mixed Facade** 混合立面

Materials 材料 | Lightweight Aluminum Composite Panels, Precast Concrete Panels 轻质复合铝板、预制混凝土墙板

Detail 细部

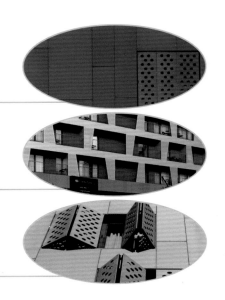

Location: Tuheljske Toplice, Croatia
Architects: MVA / Marin Mikelić, Tomislav Vreš
Project Area: 12,250 m²
Stories: Basement + ground floor + 4 stories
Photography: Ivan Dorotić

项目地点：克罗地亚 Tuheljske Toplice
建筑设计：克罗地亚 MVA 建筑师事务所 / Marin Mikelić, Tomislav Vreš
项目面积：12 250 m²
楼层：地下室 + 地面层 +4 层
摄影：Ivan Dorotić

Hotel Well
Well 酒店

Features 项目亮点

The selection of materials (lightweight aluminum composite panels and precast concrete panels) for the facade envelope marks the theme of duality (light / heavy); this intervention aims to establish new and more complex spatial relations of the existing space.

立面上金色的轻质复合铝板，与预制混凝土墙板形成"轻"与"重"的对比；整个建筑的设计使新建筑与原建筑形成了新的更复杂的空间对话。

Overview 项目概况

The addition to an existing congress/wellness hotel is located in the thermal complex "Terme Tuhelj", Croatia, next to the protected historical park and baroque Curia "Mihanović". The starting point of the project was the idea of connecting all the existing and new facilities and integrating them into a new and meaningful ensemble. The concept of a "detached" hotel, providing the users a connection with nature from all the inside spaces.

Designed by Marin Mikelić and Tomislav Vreš of MVA architecture, the program of the new part of a hotel is clearly vertically faparated (public + accommodation). 126 rooms are divided into three smaller pavilions (along to the three old ones) and the "public" facilities are organized into elongated ground base.

这是一座现有的国会保健酒店的扩建项目，位于克罗地亚，靠近历史保护公园和巴洛克风格的元老院。设计目标是要将所有保留建筑与新建筑整合在一起，形成一个新的整体，因此设计师构思了一个"分离"式的酒店，让使用者在室内的每一个地方都能亲近自然。

这座建筑由 MVA architecture 设计事务所的 Marin Mikelić 和 Tomislav Vreš 设计，公共空间与客房在竖向维度上是完全分开的。126 套房间分散在 3 座建筑中（与三座旧建筑一起），公共空间都设置在狭长的地下室中。

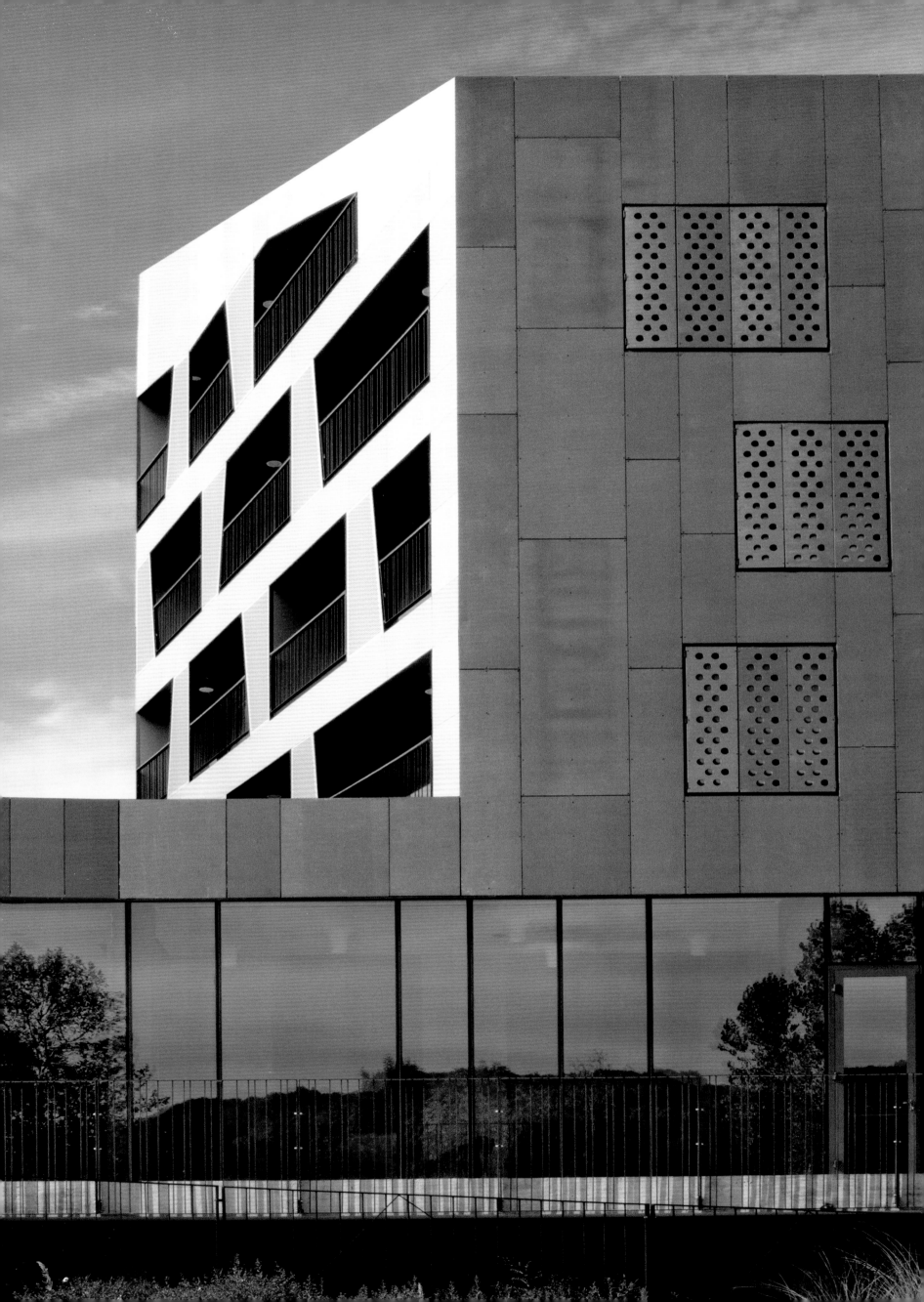

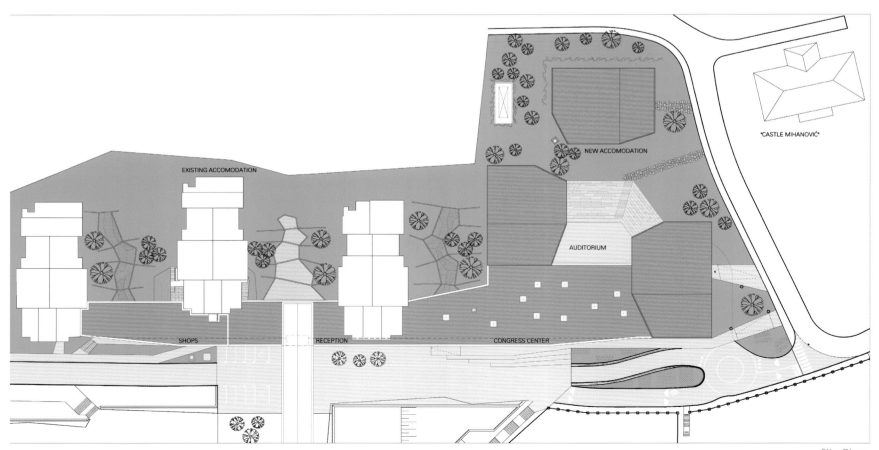

Site Plan

Typical Floor Plan

Facade and Materials 立面与材料

The selection of materials for the facade envelope marks the theme of duality (light / heavy). The exterior siding is "wrapped" with the envelope of lightweight aluminum composite panels coloured in gold whose perforations allow views from the rooms. The rest of the facades and roof surfaces are covered with precast concrete panels, partly opening as perforated 'eyelids' in front of the windows. This intervention aims to establish new and more complex spatial relations of the existing space.

　　表皮材料的选择体现了"轻"与"重"的对比。外部的滑轨窗使用了金色的轻质复合铝板，上面的圆孔使得从室内能够看到外部的景观，从外部却很难看到室内。立面上的其他部分和屋顶都采用了预制混凝土墙板，有些可开启的穿孔预制板设置在窗前，功能像是眼睑一样。整个建筑的设计使新建筑与原建筑形成了新的更复杂的空间对话。

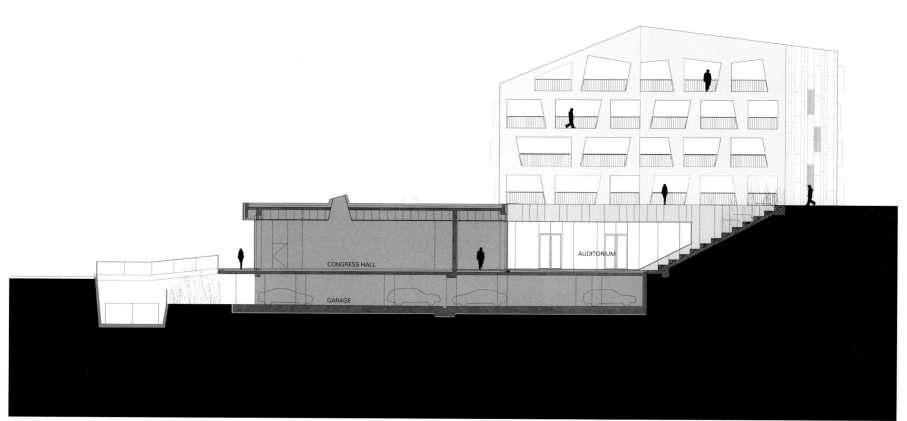

Section

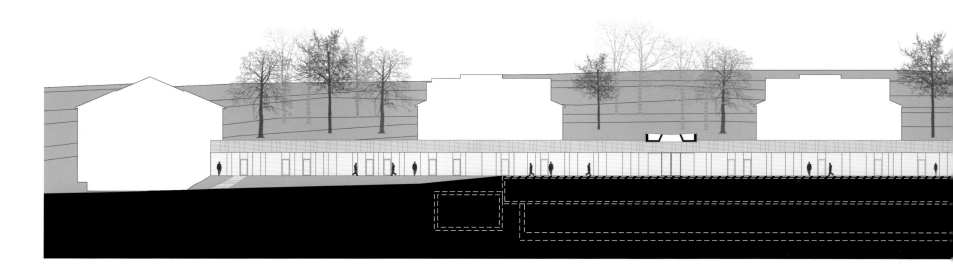

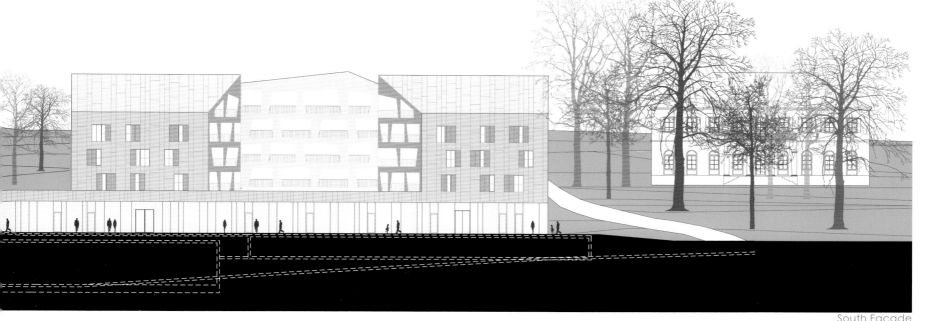

South Facade

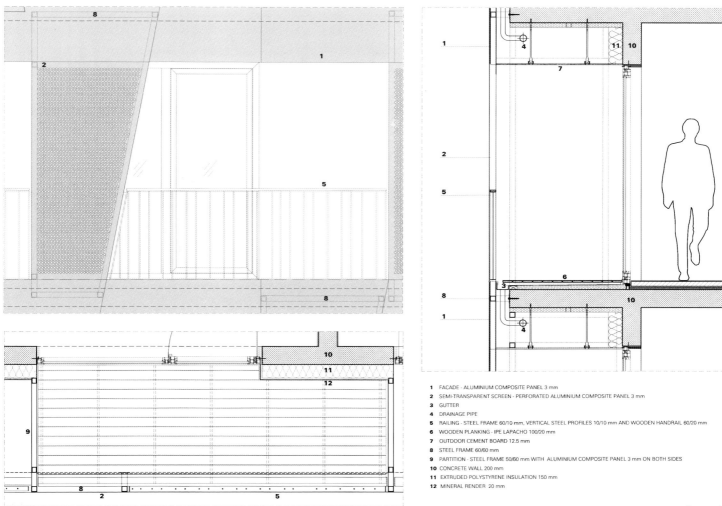

1. FACADE - ALUMINIUM COMPOSITE PANEL 3 mm
2. SEMI-TRANSPARENT SCREEN - PERFORATED ALUMINIUM COMPOSITE PANEL 3 mm
3. GUTTER
4. DRAINAGE PIPE
5. RAILING - STEEL FRAME 60/10 mm, VERTICAL STEEL PROFILES 10/10 mm AND WOODEN HANDRAIL 60/20 mm
6. WOODEN PLANKING - IPE LAPACHO 100/20 mm
7. OUTDOOR CEMENT BOARD 12.5 mm
8. STEEL FRAME 60/60 mm
9. PARTITION - STEEL FRAME 50/60 mm WITH ALUMINIUM COMPOSITE PANEL 3 mm ON BOTH SIDES
10. CONCRETE WALL 200 mm
11. EXTRUDED POLYSTYRENE INSULATION 150 mm
12. MINERAL RENDER 20 mm

Facade Detail

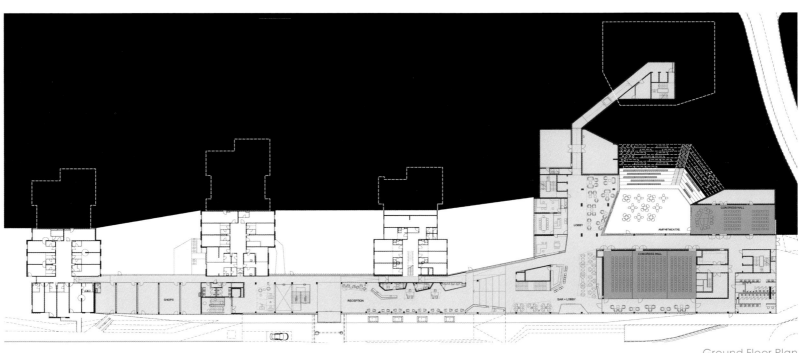

Ground Floor Plan

233

KEYWORD 关键词	Skin and Facade 表皮与立面	Mixed Facade 混合立面
	Materials 材料	Glass, Concrete 玻璃、混凝土
	Detail 细部	

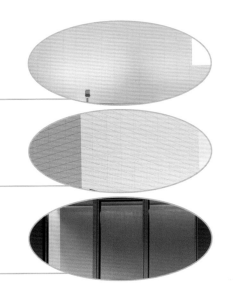

Location: Mengibar, Spain
Developer: Ogamar S.A.
Designer: DTR_STUDIO (Jose Maria Olmedo+Jose Miguel Vazquez)
Collaborators: Emilia Coronado Y Claudia Gutierrez, Tomas Pancorbo, Miliario Engineering
Land Area: 2,300 m²
Photography: Javier Callejas

项目地点：西班牙门希瓦尔
开发商：Ogamar S.A.
设计：西班牙DTR事务所（Jose Maria Olmedo，Jose Miguel Vazquez）
合作团队：Emilia Coronado Y Claudia Gutierrez, Tomas Pancorbo，Miliario Engineering
占地面积：2300 m²
摄影：Javier Callejas

Hotel in Mengibar
门希瓦尔酒店

Features 项目亮点

The architects set a powerful front towards the road and releases all existing garden area and relaxation area and expansion of the hotel.

设计师为该建筑设计了强有力的正面，使酒店与周边的园林区、休闲区和谐相融。

Overview 项目概况

The Hotel, as a unique and reference must be shown and positioned in a prominent place. Thus, the placement of the same area of the parcel abutting the highway Bailen brings us rolled flows both from the industrial zone as the highway exits.

The local Mandatory Regulations lead the architecture to a forced separation from boundaries of four meters, maximum height 9.60 meters and a floor area of no more than 1,000 m². The designers establishes a path from the clients out of the car with suitcases, until it leaves the luggage to their rooms, after they enter into slight ramp to locate the hotel on a podium that distinguishes it from the rest.

酒店作为一个独特的存在个体，优越的项目位置是不可或缺的。项目所在的工业区毗邻拜伦高速公路，高速入口人来人往。

根据当地法规的强制规定，项目建筑必须离边界4m，最高不得超过9.6m，建筑面积不得超过1 000 m²。酒店坐落于一座小斜坡上，与周边其他建筑明显区分开来。设计师为顾客建立一条通道，方便带着旅行箱下车的客户直达酒店房间。

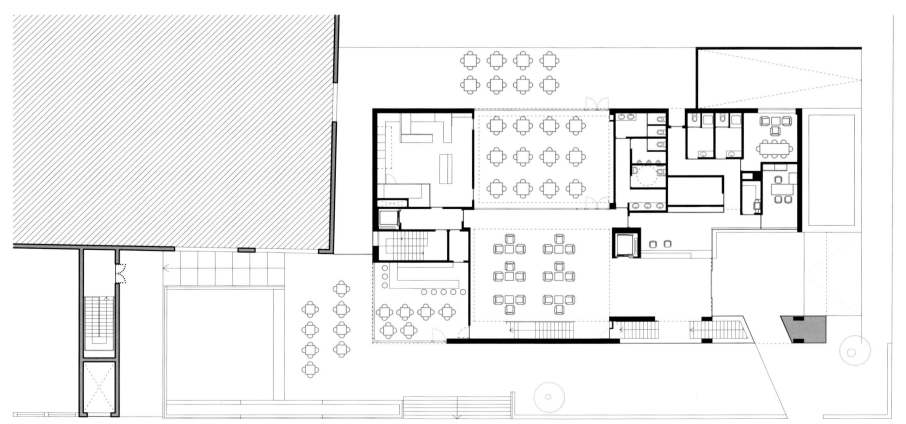

02 Ground Floor

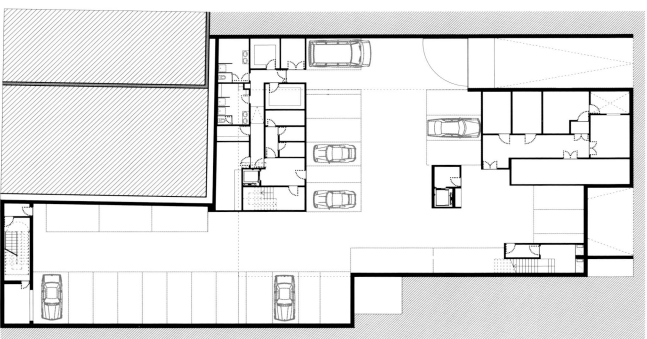

01 Basement Floor

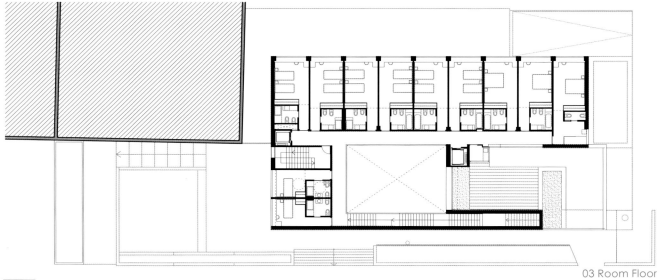

03 Room Floor

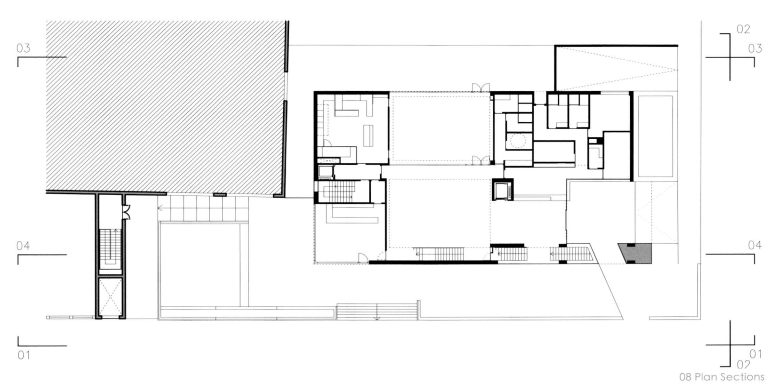

08 Plan Sections

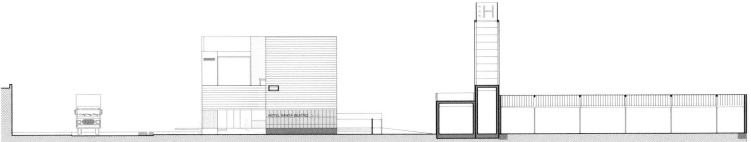

05 Section 02

Skin and Facade   表皮与立面

This position leads the architects to create a powerful front towards the road and releases all existing garden area and relaxation area and expansion of the hotel. In addition they make an enclosure that appropriates new existing ships giving them a new character that will be enhanced with subsequent conversion into halls for weddings.

Great reception vestibule, low ceilings leading the traveler to go up to the hall to triple height that is piece of ideal world that generates the hotel itself to forget their surroundings and staring at those points of interest: the garden through the conservatory restaurant and sky through the large window situated north. This space is enriched with circulations that occur around them, either by the two flights of stairs to the lobby to open up to the rooms, either by entering the halls of the dorms.

　　针对项目位置的特殊性，设计师为建筑设计了强有力的正面，使酒店的扩张与现有的所有园林区、休闲区同步和谐。设计师为酒店添加了围墙，赋予其新的特色，将其改造成为婚礼大堂。

　　宽敞的接待前厅与低矮的天花板在视觉上增加了大厅的高度，旅行者可以在这里感受理想的世界，忘记周边的环境。体验以下美景：贯穿温室餐厅的花园，透过北面大窗户所看到的天空。大厅周围的设计丰富了空间形式，比如从大堂到各个房间的之间的两层楼梯，或者从大厅进入的员工宿舍。

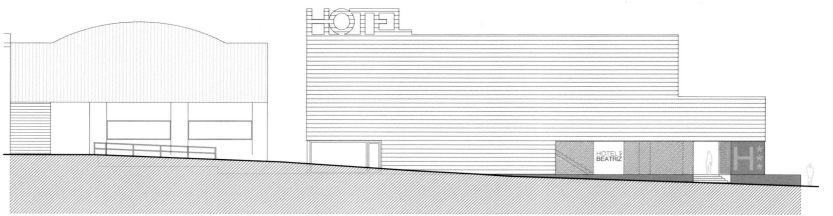

Section 01

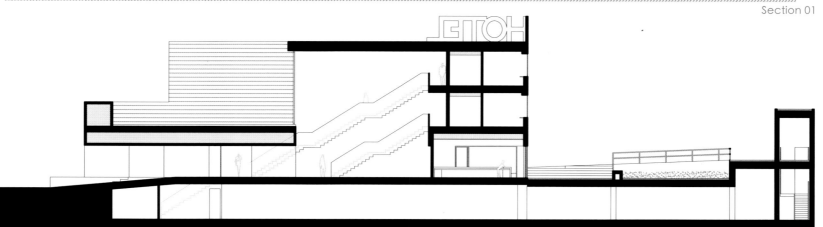

Section 04

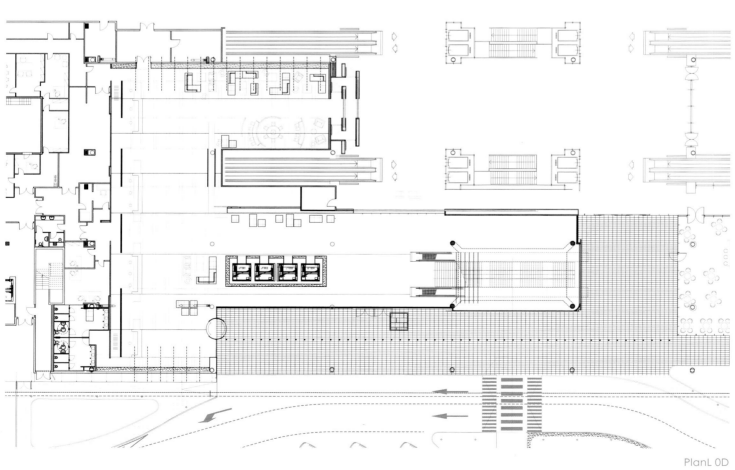

PlanL 0D

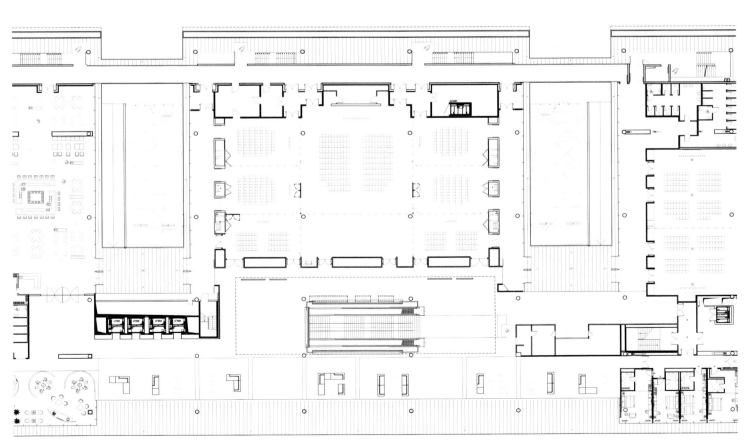

PlanL1D

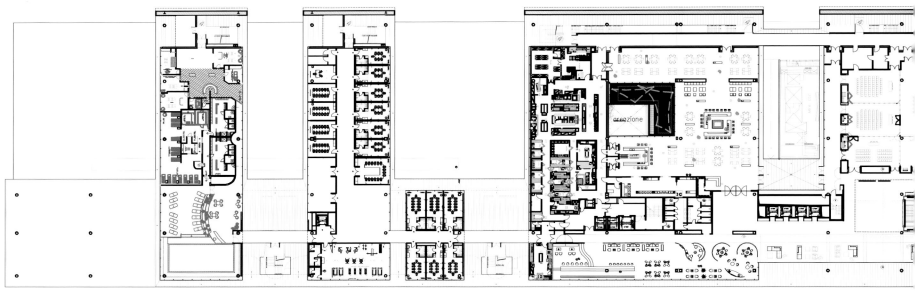

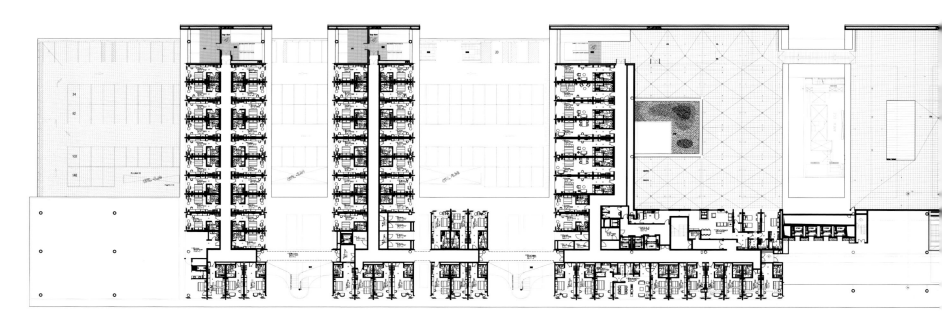

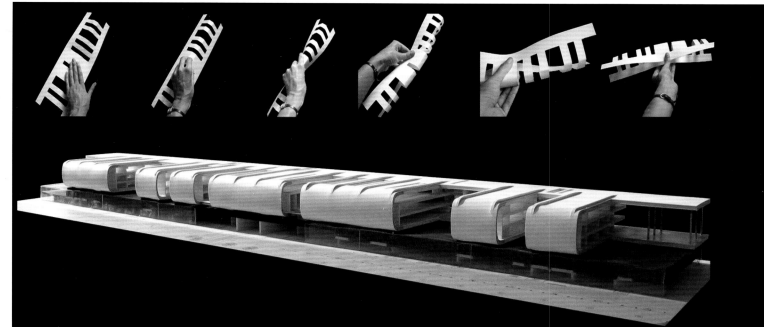
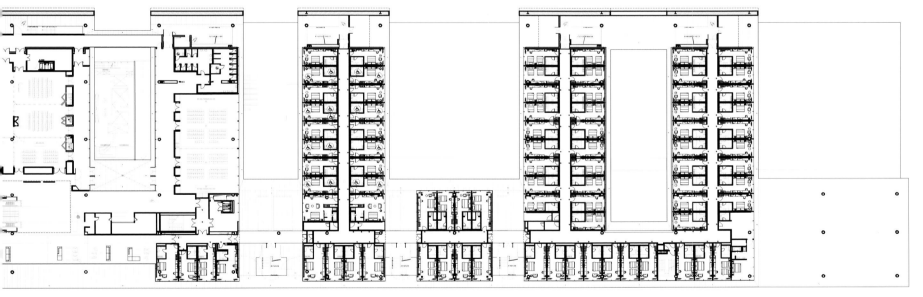

Plan L1

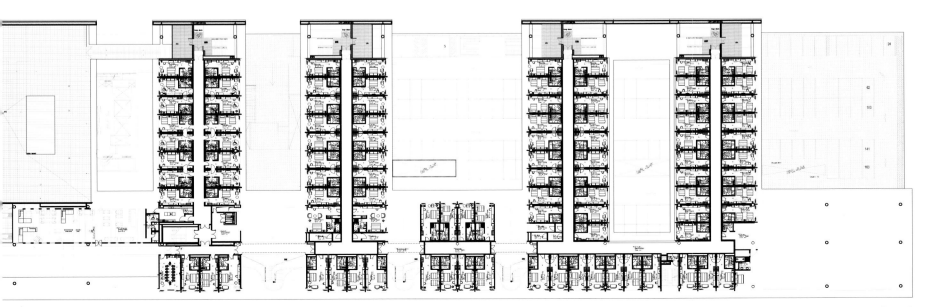

Plan L3

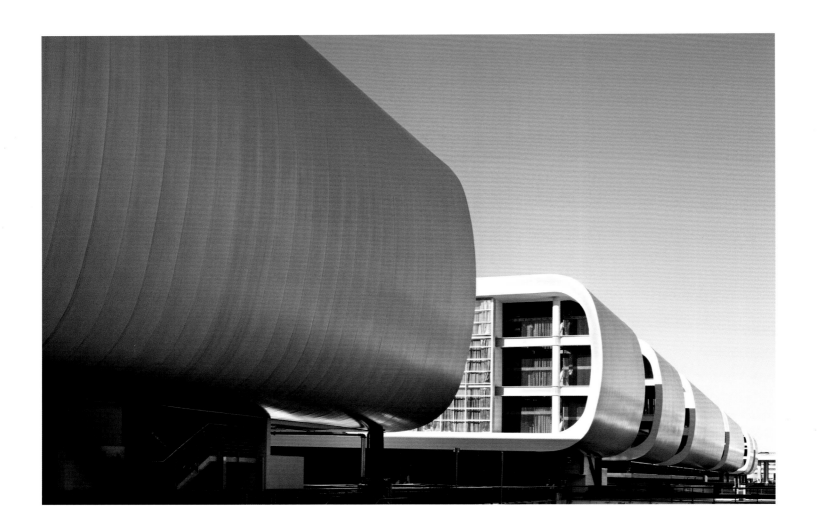

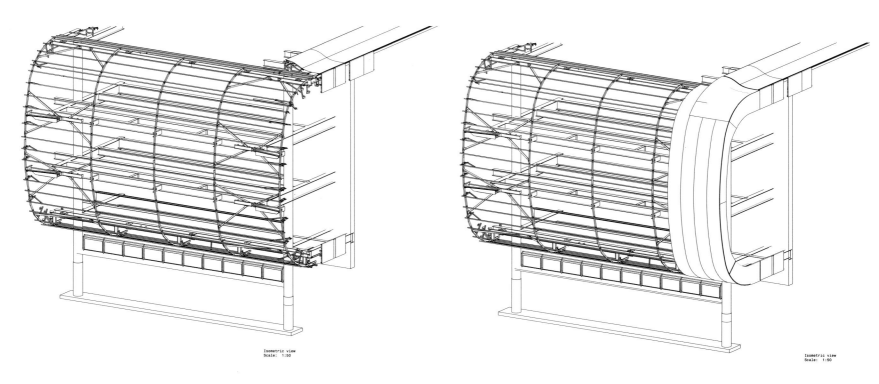

Isometric view
Scale: 1:50

Isometric view
Scale: 1:50

Facade and Materials 立面与材料

Designers were looking for a seamless shell to fold around the functional volumes of the buildings. After researching and detailing a number of alternatives they finally opted for pultruded fibreglass panels. The material has a series of qualities highly suitable for building: it is light weight, elastic, very stable in extreme temperatures (-20°C to +50°C), fireproof and waterproof. The shell is curved around a subframe of pultruded structural profiles and steel arches, wrapping the entire length of each bay, or module of the hotel in a smooth, continuous skin. The length of the hotel comprises seven bays separated by courtyards, with the curved façades facing the airport, and the rooms looking out on the courtyards.

The movement of people as they arrive and leave the airport, the tension created between solid and void, curved and straight line, the play of light reflected off and through the building, provide a variation of views and give a dynamic to the architecture that designers were looking for.

建筑外立面采用拉挤纤维玻璃板，这种材料有以下特点：轻量、弹性好、稳定牢固（即使是在 -20°C 和 +50°C 等极端温度的情况下）、防火、防水。这些纤维玻璃板形成的外壳包裹着建筑，形成了一面光滑连续的外壁。酒店共由七个部分（隔间）组成，曲线的一面面对机场，庭院穿插其中，客房的窗户朝庭院而开。将七个隔间串联起来的部分由拉挤型材的结构剖面与钢拱组成。

往来机场的人可以看到西立面不规则的虚实设计，配合灯光效果，产生一种深度感和动态感，这正是设计师所期待的。

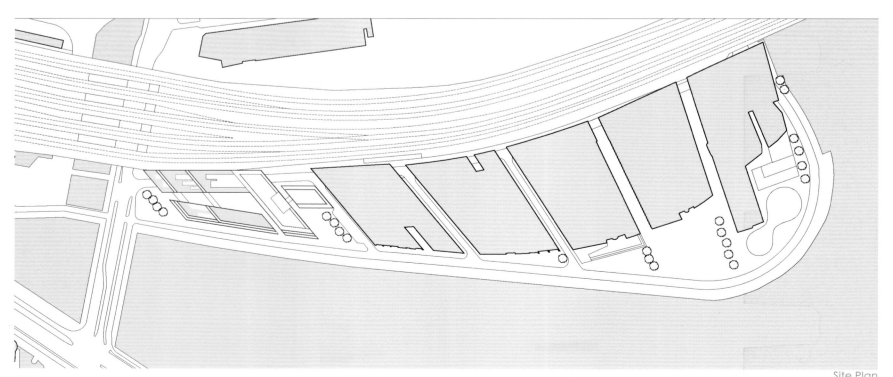

Site Plan

Structure and Shape 结构与造型

In response to the masterplan, the hotel is arranged around a courtyard, with three linked blocks of bedroom accommodation rising above two public floors containing conference and dining facilities. The building's composition ensures that the bedrooms, terraces and Sky Lounge make the most of the spectacular views across the water and the old town. The large bedroom suites, which sit in the angular corners of the building, enjoy panoramic views across the dock and the central courtyard allows high levels of daylight into the bedrooms. The public areas spill out onto a sunny pedestrian quayside facing the city: on the north side, the island's access road undercuts the building and produces a cantilevered range of accommodation that accentuates its angular plan.

根据总体规划设计,酒店设置在一个庭院四周,卧室区在两层公共区上,公共区包含会议和餐饮设施。建筑的结构保证了卧室、露台和空中走廊都可以欣赏到城市大部分的壮观景色。坐在角落的大套房,可以欣赏到码头全景,中央庭院的设置有助于更多阳光进入卧室。公共区延伸到一个阳光充足的行人码头:北面,通往岛上的路镶嵌在建筑底边上,形成一个悬臂调节幅度,突出了建筑的棱角。

Facade and Materials　　立面与材料

The facades consist of moveable perforated shutters, glazing and brickwork, emulating the spirit of traditional Dutch architecture. The palette of materials in each part of the building responds to the varying architectural styles which surround it. The south-facing elevation, with its silver blue brickwork responds to the vernacular of the old town it faces, while the glazing on the east and west facades and bridges between the three blocks reflects the architectural language of the more contemporary parts of the city. The internal courtyard facades are clad in red zinc, responding to the changing quality of light.

　　外立面由可移动的穿孔百叶窗、玻璃和砌砖构成，沿袭了传统的荷兰建筑精神。建筑每个部分材料的颜色都与周围不同的建筑风格相呼应。南向的立面所用的银蓝色砌砖彰显着本地的特色，而东西方向上的玻璃外立面和连接三个街区的桥梁反映了这个城市更现代的建筑语言。内部庭院外立面镀上了红色的锌，可以应对不同的光质。

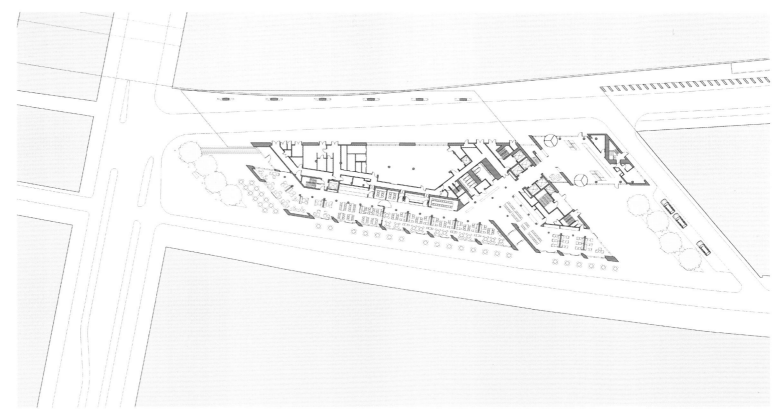

Ground Floor Plan

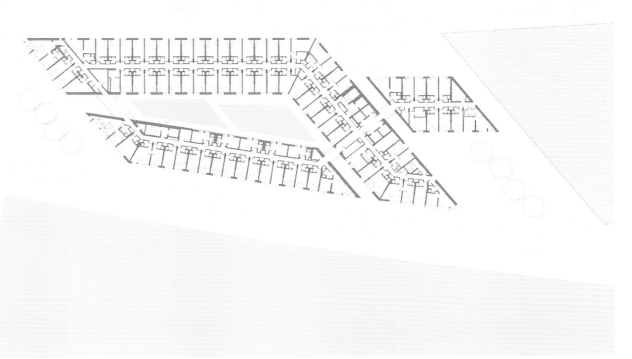

Typical Floor Plan

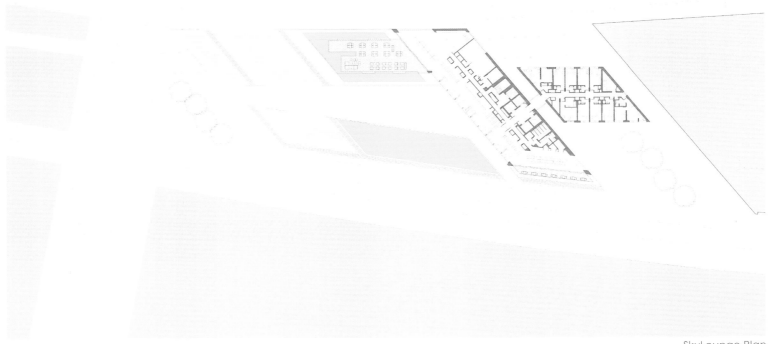

SkyLounge Plan

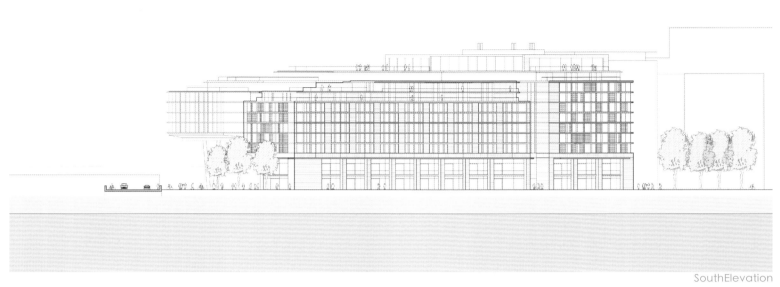

South Elevation

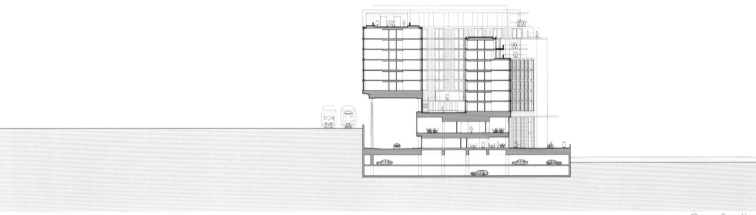

Cross Section

Open Form
Innovative Style
Streamline Design
Interactive Space

开放形态
新颖造型
流线设计
互动空间

Transportation and Sports Building
交通体育建筑

Transportation buildings include the transportation hubs and transition facilities for railway, road and airline, while transportation buildings are the facilities for sports education, competition, physical exercise or sports entertainment. These two categories are important public infrastructures in our daily life. With the development of urban transportation and sports activities, buildings of these types have also developed and usually impressed us with unique appearance, special structure or specific functions.

In terms of architectural design, transportation buildings and sports buildings have their similarities and also characteristics. For the building form and appearance, it usually pays attention to modern sense and aesthetic feeling by using glass facade, light metal facade or steel facade as well as reinforced concrete structure. And for the space design, it emphasizes openness and transparent sense. Transportation building pays much attention to people flows, vertical traffic and safety, while sports building emphasizes convenient location and the connection with green parks. Since these two categories are part of public infrastructures, their artistic quality and functions should also be carefully considered in design.

交通建筑包括铁路、公路、航空交通枢纽以及交接转换设施等建筑，而体育建筑则是作为体育教育、竞技运动、身体锻炼和体育娱乐等活动之用的建筑，两者均是生活中重要的公建设施。随着城市交通以及体育事业的发展，交通体育建筑也呈现出快速发展的态势，其非常规的建筑造型、特殊的建筑结构以及特定的功能形态常常令人印象深刻。

在建筑设计上面，交通、体育建筑既有相似之处又有各自的特点，在形态和造型上往往强调时代感美感，立面以玻璃、轻质金属和钢材等作为主要的材料，建筑结构通常为钢筋混凝土结构；空间设计上具有很强的开放性和通透感。交通建筑要考虑客流、垂直交通、安全等问题；而体育建筑则布点均衡、位置适中、交通方便，并尽量与公园绿地相结合，采用集中式或者分散式布局。同时，交通体育建筑承担着公共建筑的职能，建筑的艺术性与功能性也是设计重点考虑的问题。

KEYWORD 关键词	Skin and Facade 表皮与立面	Metal Facade 金属立面
	Materials 材料	Steel, Glass, Polycarbonate 钢铁、玻璃、聚碳酸酯等
	Sustainable Building 可持续性建筑	

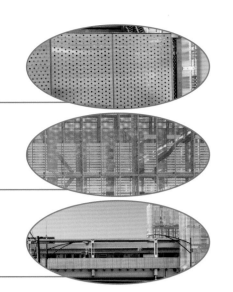

Location: Chicago, USA
Client: Chicago Department of Transportation
Designer: Ross Barney Architects
Photography: Kate Joyce Studios

项目地点：美国芝加哥
客户：芝加哥交通局
建筑设计：Ross Barney Architects
摄影：凯特·乔伊斯工作室

CTA Morgan Street Station
CTA 摩根车站

Features 项目亮点

To meet the requirements of sustainable development, the design took the neighborhood environment as reference to choose materials or layout and made itself the golden model of the modern urban rail transit system in Chicago.

设计从可持续发展的要求出发，无论是材料的选择还是车站的布局，均以车站周边的环境作为参照物，堪称芝加哥现代城市轨道交通体系的典范。

Overview 项目概况

The new Morgan Street station defines the geographic center of the Fulton Market District, with its rich combination of warehouses, industrial loft conversions, restaurants, specialty food purveyors, and boutique stores. This new gateway to the Chicago Loop serves as a strong emblem of the modernity of Chicago's mass transit system.

CTA 摩根车站位于富尔顿市场区域的中心。这片区内聚集了众多大大小小的仓库、旧楼改造而成的工作室、餐厅、特色食品店和精品店。新车站是芝加哥现代化城市轨道交通系统的一个极好的案本。

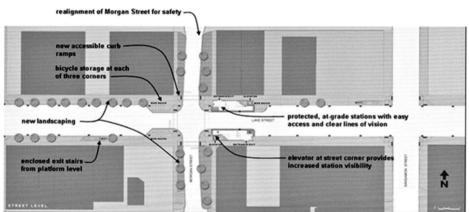

ground level site plan

Structure and Materials 结构与材料

Material cues from the neighborhood; steel, glass, concrete, polycarbonate, granite and cast iron, were chosen to reinforce the feeling of openness. Canopies above the platform are constructed of lightweight, translucent polycarbonate panels that provide weather protection, permit natural light to reach the platform, and allow for less canopy structure and ease of replacement.

Sustainability was achieved through the use of high amounts of recycled content in steel and concrete. Polycarbonate panels have a high recycled content and are regionally produced; granite flooring was extracted from regional quarries; and glazing was regionally produced.

Project landscaping is drought tolerant, requires no irrigation and minimizes storm water runoff. New bicycle racks encourage the use of alternative transportation.

建筑材料选择从车站周边环境获得灵感。钢铁、玻璃、混凝土、聚碳酸酯、花岗岩、铸铁的使用增强了开放的体验。站台天篷用的是聚碳酸酯板，重量轻，半透明状不仅能有效档雨，采光好并且结构简单，便于更换。

通过使用大量含钢材和混凝土的回收物，达到了建筑可持续性的要求。聚碳酸酯板回收性强，又是当地资源，花岗岩地板来自本地的采石场，玻璃也是当地生产的。

绿化用的植物耐旱，不需要灌溉，能最大限度地减少雨水流失。新的自行车停车位鼓励绿色出行。

KEYWORD 关键词	Skin and Facade 表皮与立面	Mixed Facade 混合立面
	Materials 材料	Concrete, Aluminium Plate, Glass 混凝土、铝板、玻璃
		Energy-Saving Architecture 节能建筑

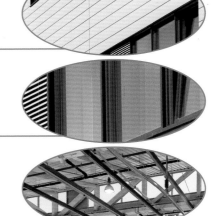

Location : Vienna, Austria
Client: ÖBB Bau AG
Designer : Zechner & Zechner ZT GmbH
Site Area: 36,000 m²
Construction Area: 25,160 m² (covered area), 35,530 m² (usable area)
Photography:Thilo Härdtlein, München

项目地点：奥地利维也纳
客户：ÖBB Bau AG
建筑设计：奥地利 Zechner & Zechner 设计公司
占地面积：36 000 m²
建筑面积：25 160 m²（覆盖面积），35 530 m²（实用面积）
摄影：Thilo Härdtlein, München

ÖBB Rail Service Center
OBB 铁路服务中心

Features 项目亮点

The project is a sustainable "slim building" based on the terrain characteristics, with various types of materials in means of technology.

设计很好地把握地形的特征，运用多种材料和技术手段，建造了一个可持续发展的"苗条建筑"。

Overview 项目概况

After an EU wide competition the Austrian Federal Railways ÖBB has commissioned "Wiener Team", a consortium of planning consultants from architecture and structural design, structural planning, load bearing structural planning, traffic planning and HVACR planning for the overall planning of the new Vienna Central Train Station. Part of the main project is the construction of a service centre on the site of the former Matzleinsdorf freight terminal in the vicinity of the future central station. The architects Zechner & Zechner are responsible for the architectural design of the facility.

通过欧洲范围内的竞赛，奥地利联邦铁路公司 OBB 委托了"维也纳团队"担任新维也纳中央火车站的整体规划项目顾问，包括建筑和结构设计、结构规划、承重结构规划、交通规划、供热通风与空调工程规划。本案正是其中的主体工程之一。在未来的中央车站附近，前 Matzleinsdorf 货运站建设一个服务中心，由 Zechner & Zechner 设计公司负责该设施的建筑设计。

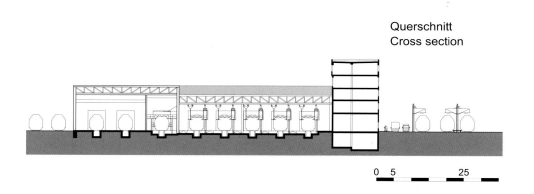

Querschnitt
Cross section

0 5   25

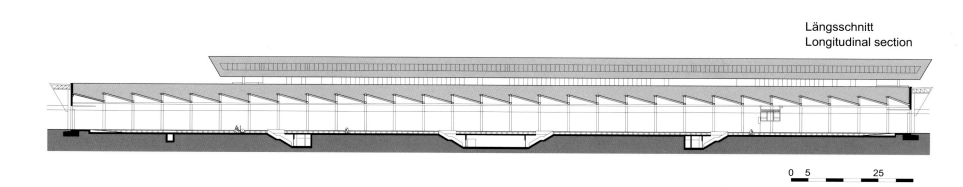

Längsschnitt
Longitudinal section

0 5   25

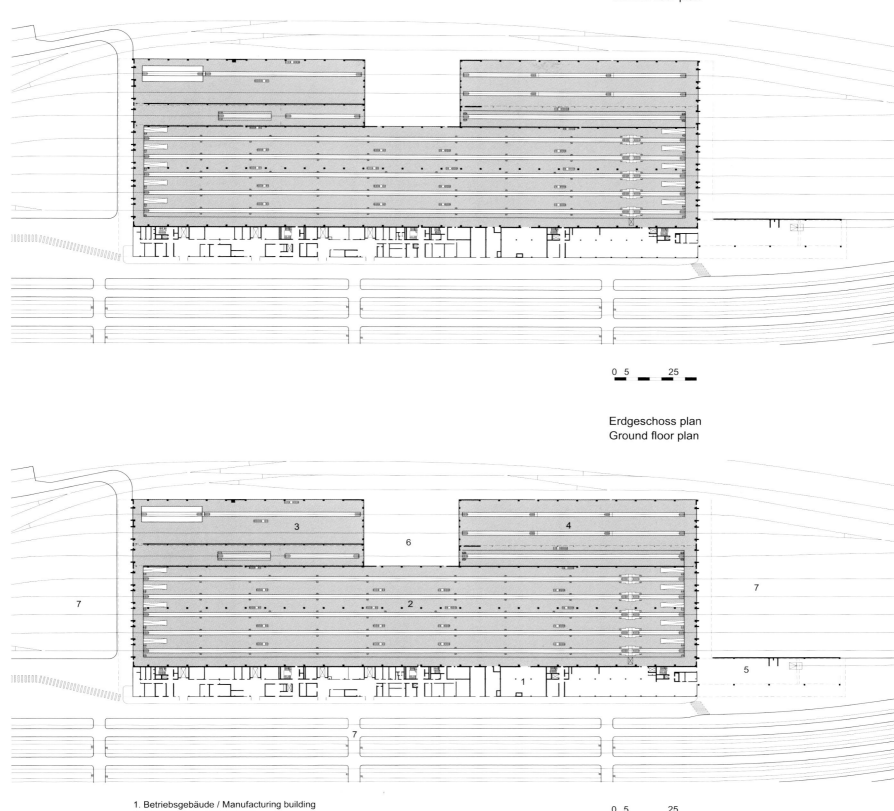

Erdgeschoss plan
Ground floor plan

Erdgeschoss plan
Ground floor plan

1. Betriebsgebäude / Manufacturing building
2. Blockzughalle / Train block
3. Traktionshalle / Towage block
4. Lok&Wagenhalle / Loco block
5. Freilager / Open storage
6. Schiebebühne / Traverser
7. Gleisbereich offen / Track area open

Shape and Structure 造型与结构

The goal of the construction work is the bringing together of servicing, maintenance and cleaning facilities for block trains, passenger train wagons and locomotives at the site. The spatial concentration of facilities for ÖBB technical services, ÖBB traction and ÖBB passenger transport produces increases in efficiency in servicing, vehicle maintenance and availability, and in this way reduces maintenance costs.

The approx. 240m long and approx. 13.5m wide services building has 5 floors, and in essence comprises of the administration, production, storage, personnel and service areas for the ÖBB technical services, traction and passenger divisions. Furthermore the building also serves as a support centre for train catering and for a sleeping car operator.

该项目的目的是把编组列车的维修保养和清洗设施、客运火车车厢和机车联合在一起，通过OBB技术服务、牵引和客运设施的空间集聚提高服务、车辆维修和可用性的效率，从而降低维护成本。本项目的结构平面规划基于一个先进的维修理念。

服务大楼长约240m，宽约13.5m，共5层，内设OBB技术服务、OBB牵引和OBB客运部门的管理、生产、储存、人员和服务区。此外，大楼还作为列车餐饮和卧铺车厢运营商的支持中心。

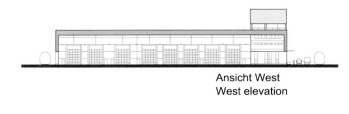

Ansicht West
West elevation

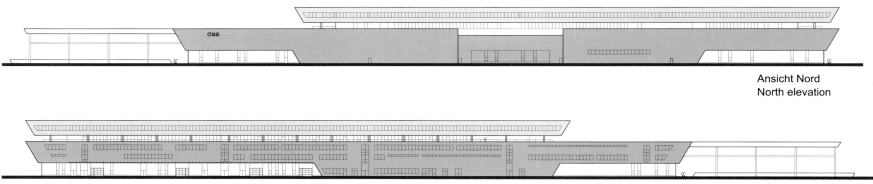

Ansicht Nord
North elevation

Ansicht Süd
South elevation

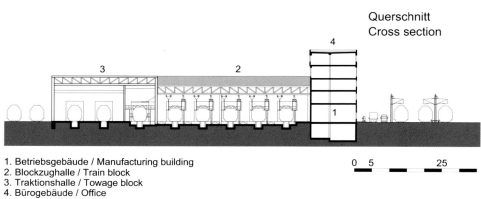

Querschnitt
Cross section

1. Betriebsgebäude / Manufacturing building
2. Blockzughalle / Train block
3. Traktionshalle / Towage block
4. Bürogebäude / Office

0 5   25

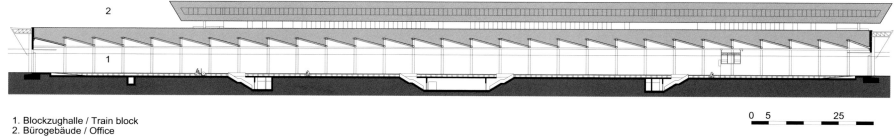

Längsschnitt
Longitudinal section

1. Blockzughalle / Train block
2. Bürogebäude / Office

0 5   25

Skin and Materials 表皮与材料

The service building is located amidst railway tracks, and therefore requires an elongated, narrow configuration. The dynamic of the passing rail traffic and the linearity of the trains are reflected in the design of the building. The materials used for the outer shell, such as exposed concrete, aluminium panels and profile glass emphasise the industrial character of the facility.

To build a sustainable "slim building" with low primary energy consumption while providing a healthy and comfortable environment, was fulfilled using the following techniques:

Increased insulation and airtightness of the building shell due to a compact design
Making use of the thermal mass
Regulatable external blinds
Optimum use of transparent facade areas
Natural ventilation, heat dissipation via ventilation at night
Systematic use of waste heat
Efficient illumination technology with daylight-dependant control

服务大楼位于铁路轨道之间，这也决定了其细长狭窄的体量特征，其设计还反映了往来的轨道交通的活跃以及列车的线性特征。建筑外壳的材料如裸露的混凝土、铝板和浮雕玻璃都强调了这个建筑设施的工业特征。

为了建设一个可持续发展的"苗条建筑"，一次能源消耗低且提供一个舒适健康的环境，需要使用以下技术：建筑外壳采用紧密设计以增加保温和气密性；利用地热容量；可调控的外部百叶窗；透明立面领域的最佳应用；采用自然通风，夜间通风散热；余热的系统利用；利用日光控制的高效照明技术。

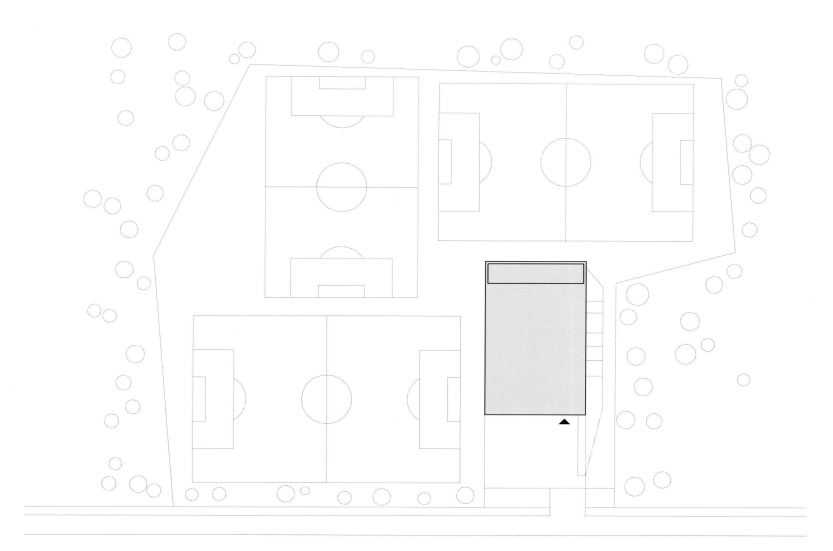

site plan lageplan

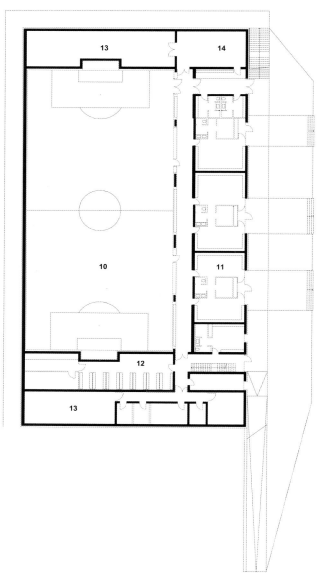

10 soccer hall fußballhalle
11 changing room umkleidekabine
12 laundry room waschküche
13 utility room technik
14 storage lager

basement untergeschoß

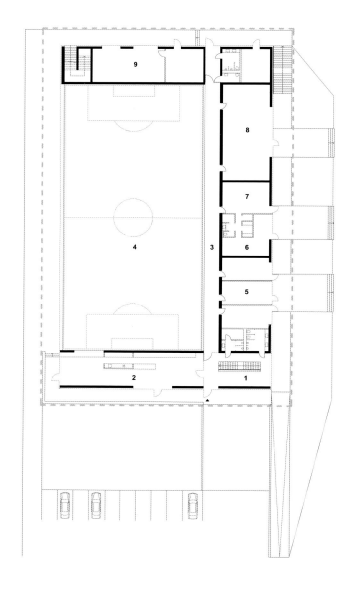

1 foyer foyer
2 canteen kantine
3 gallery galerie
4 void soccer hall luftraum fußballhalle
5 office büro
6 sauna, relaxation room sauna, ruheraum
7 massage massage
8 fitness fitness
9 storage lager

ground floor erdgeschoß

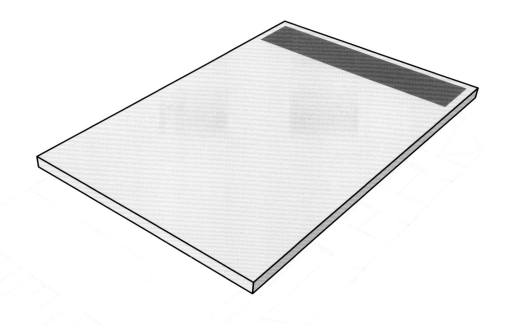

roof with terrace

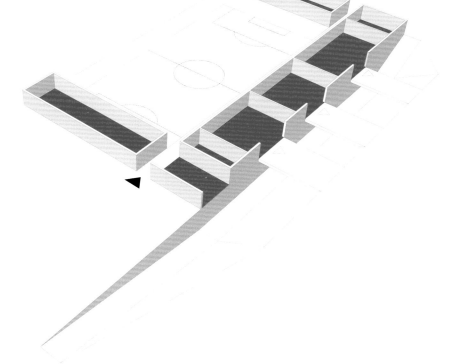

offices, fitness

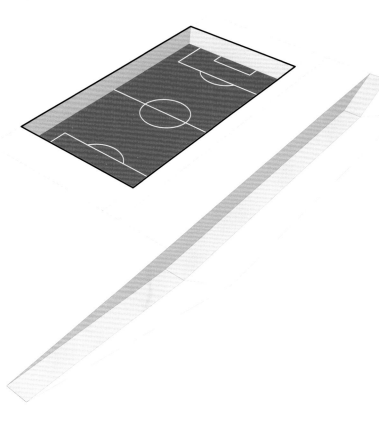

soccer hall

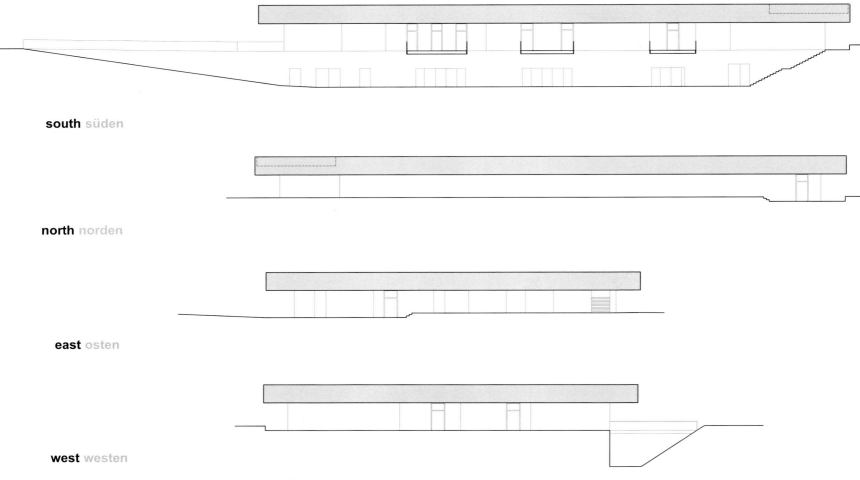

south süden

north norden

east osten

west westen

elevation ansicht

Structure and Shape 结构与造型

Vienna building regulations set a maximum building height of 4.5 m and so the 7.0 m required clearance height in the training hall created an inevitable consequence: the entire building had to be lowered half below ground. The roof cantilevers out a few meters on all sides and forms a generous, weather-resistant perimeter for entrances and terraces. On the eastern wing, a compact roof construction over the storage spaces is utilized as a roof terrace, where coaches and visitors are afforded a clear view over the heated artificial pitch.

维也纳建筑法规规定，建筑高度最高为4.5m，所以为了实现训练厅7m的净空高度，设计师将整个大楼半沉下地面。训练厅四周的屋顶悬臂结构为入口和露台挡风遮雨，建筑东侧储藏空间上方的屋顶被设计成露台，教练和观众可以在此观看人工训练场上的激烈情景。

KEYWORD 关键词	Skin and Facade 表皮与立面	Mixed Facade 混合立面
	Materials 材料	Steel, Concrete 钢、混凝土
	Structure 结构	Light Steel Structure 轻型钢结构

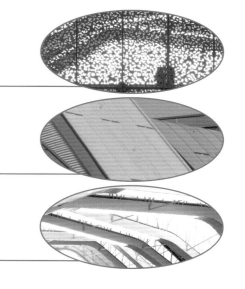

Location: Medellín, Colombia
Client: INDER
Architects: Mazzanti Arquitectos(Giancarlo Mazzanti)
+ Plan: b Arquitectos (Felipe Mesa)
Area : 30,694 m²
Photographs: Iwan Baan

项目地点：哥伦比亚梅德林
客户： INDER
建筑设计：哥伦比亚 Mazzanti 建筑设计事务所
（Giancarlo Mazzanti）
哥伦比亚 Plan: b 建筑设计事务所（Felipe Mesa）
面积： 30 694 m²
摄影： Iwan Baan

Four Sport Scenarios
四个体育场

Features 项目亮点

The steel structure and geometrical roofing system with parallel strips of the main building bodies have created an open sports space, whose perforated façades lead to a single and perpendicular bioclimatic diagram.

体育场主体采用钢筋结构，屋面通过几何平行条状设计，营造了一个开放性的运动空间，多孔立面的形态勾勒出一个垂直生物气候图。

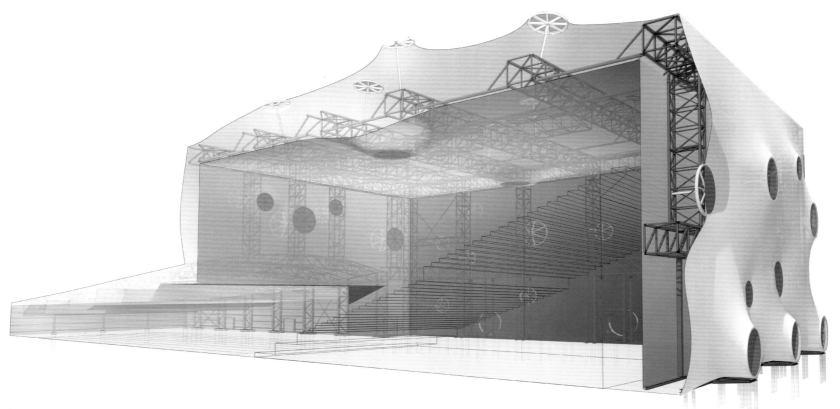

Conceptual Sketch / Magma Architecture

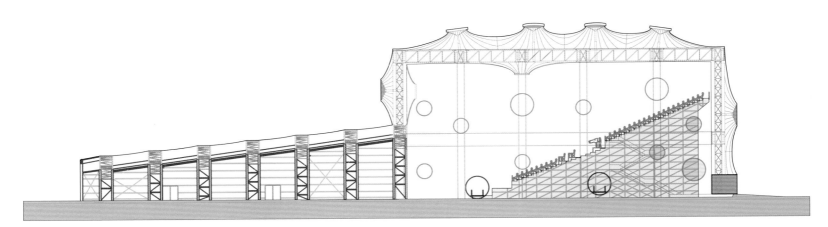

Section Finals Range / Magma Architecture

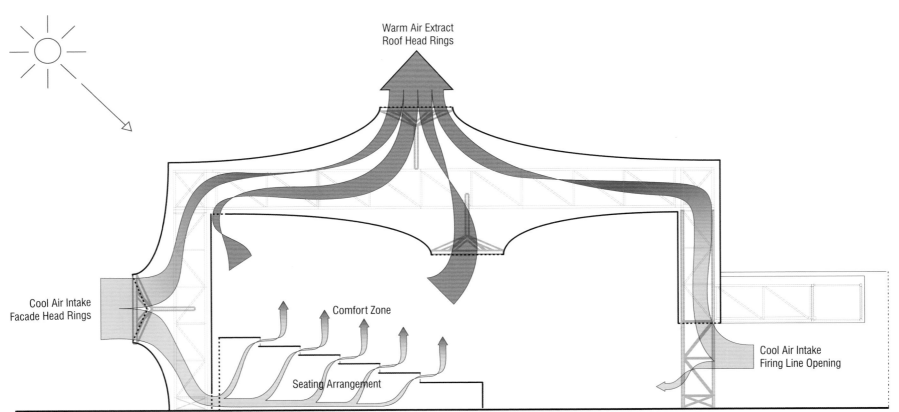

Natural Ventilation Diagram / Magma Architecture

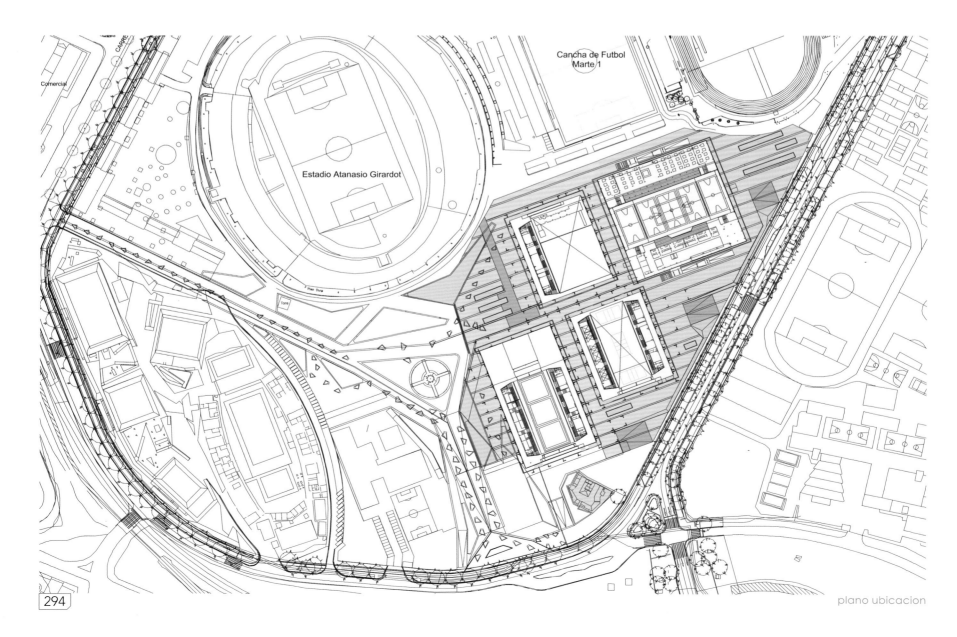

plano ubicacion

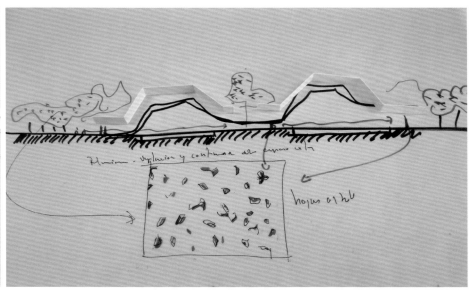
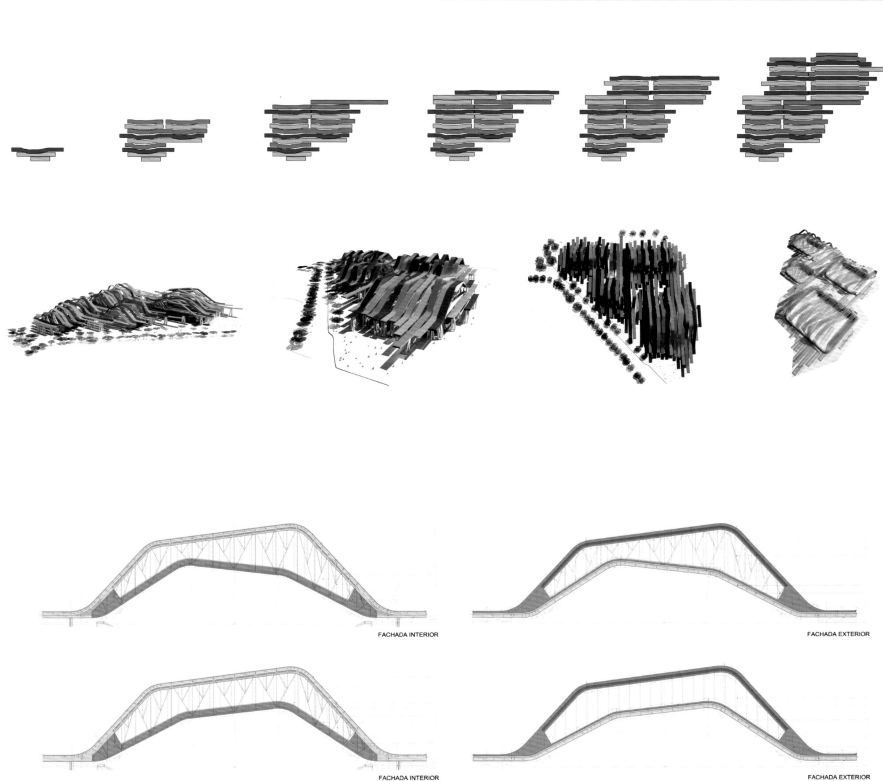

FACHADA INTERIOR	FACHADA EXTERIOR
FACHADA INTERIOR	FACHADA EXTERIOR
FACHADA INTERIOR	FACHADA EXTERIOR

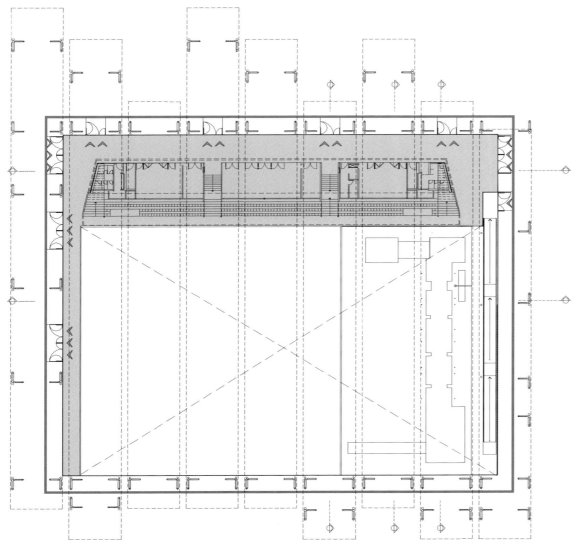

Planta Acceso Gimnasia

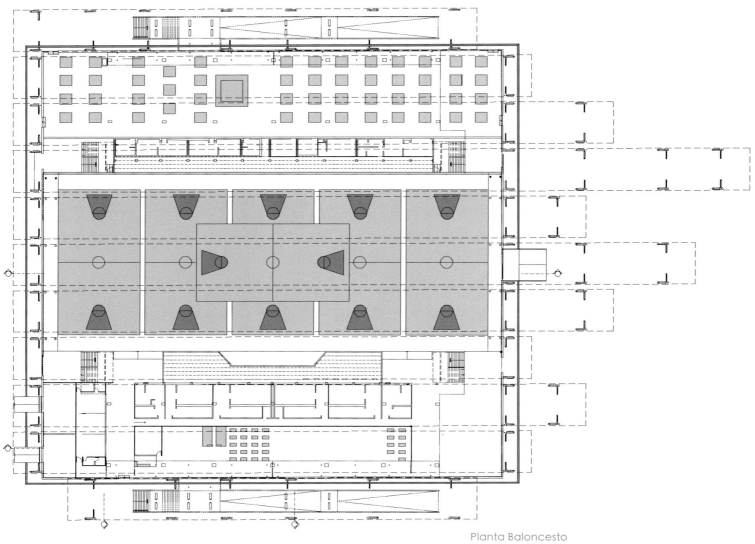

Planta Baloncesto

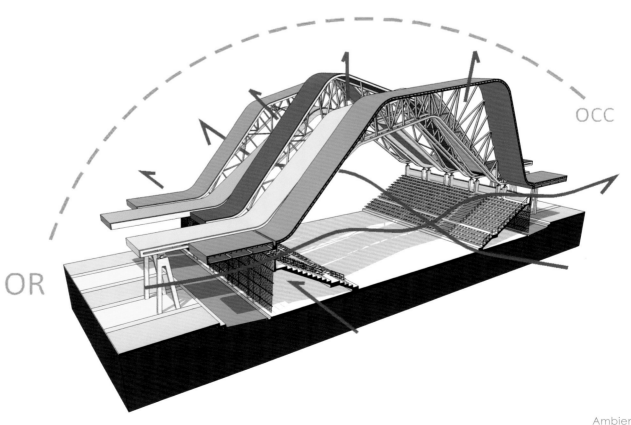

Ambiental1

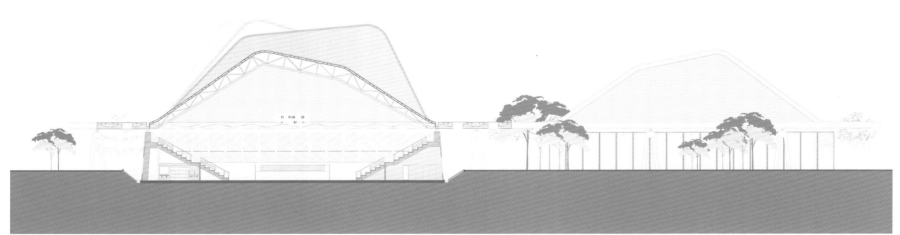

Longitudinal

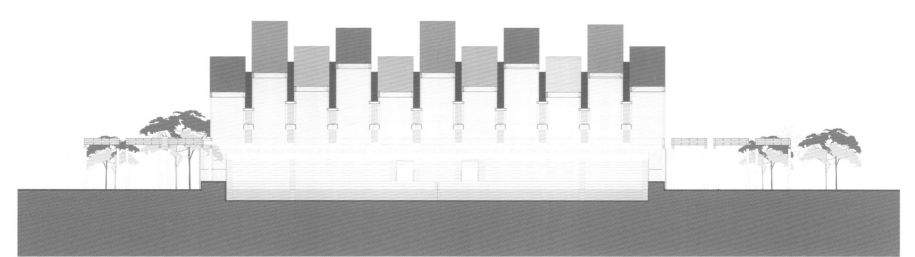

Trasnversal

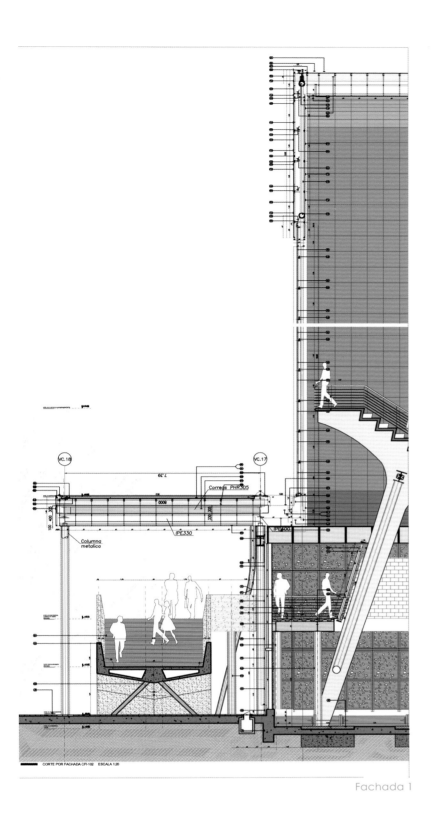

Fachada 1

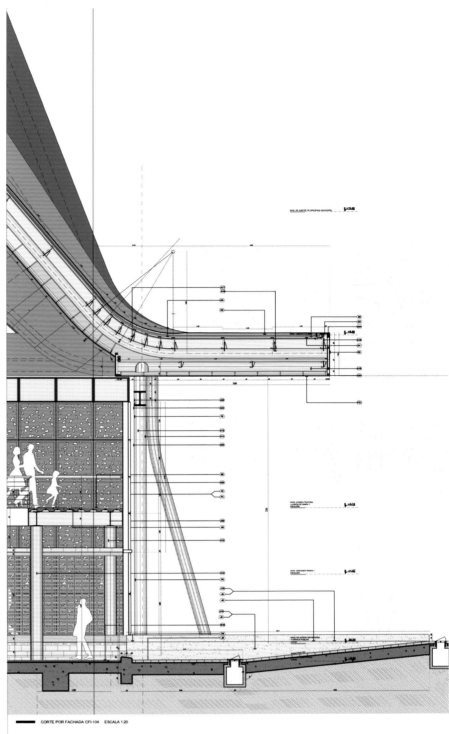

Fachada 3

Shape and Structure 造型与结构

The project has been thought as a new geography to the interior of the elongated Aburrá Valley, midway between Cerro Nutibara and Cerro El Volador. It is a building that seems to be another mountain in the city; from the remote or from the top has an abstract image geographic and festive; from the inside, the movement of the steel structure, allows the filtered sunlight to get inside the space, which is the suitable condition for the conduct of sporting events.

The project took the interior and exterior in a unified way. The outdoor public space and sporting venues are in a continuous space, thanks to a large deck built through extensive stripes out, perpendicular to the direction of the positioning of the main buildings. Each of the four sporting venues operates independently, but in terms of urban space and behave as one large continent built with public open spaces, semi-covered public spaces, and indoor sports. The solar position and the Aburrá valley winds, linked to the existing planning of the site allows, define the buildings localization. The north and south façades are open to the let the major winds pass and the east and west façades have the tribunes and eaves that control the morning and the afternoon sun. In this way they are inside of a single and perpendicular bioclimatic diagram.

项目位于细长的Aburrá山谷，Nutibara和El Volador山之间，犹如城市中的另一座山。从远处或上方看是一个抽象的地理图像，而内部的钢结构使阳光折射进入内部空间，这正是举行体育赛事的绝佳场所。

项目内外设计统一，大型甲板向外延伸突出，与主建筑位置成垂直方向，将户外的公共空间和体育场馆衔接在一起。四个体育场馆虽然采取独立设计，但整体形成一大片具有公共开放空间、半露天公共空间和室内场馆建设的城市空间。场地的规划参考了太阳的位置和Aburrá山谷的风向，以确定各个建筑的具体位置。建筑的南北立面呈开放形，便于通风；东西立面则设计了看台和屋檐以控制早晨和午后的阳光。设计师通过这样的设计在场地内勾画出一个垂直生物气候图。

Facades and Materials

The designers proposed a geometrical roofing system of parallel strips, aligned with the sun path to control it. These parallel bars act again as technical, spatial and bioclimatic patterns. They can stretch in the east-west direction, and they can be added or repeated in the north-south direction. They can also gain height to fit the different sports requirements, or go down to make a cover from the sun.

These stripes are really a 7 parallel metallic trusses system which geometry changes and relates directly with the city mountainous or topographic context. A metallic columns system, made by three assembled columns, supports the roof system and acts as a rain gutters, conducting water to the city drainage system. With a unique constructive detail, perforated façades, projecting stripes, the designers are able to solve, in reiterative ways, bioclimatic, spatial and relational aspects. The sports scenarios do not require mechanical acclimatization, using the perforated façades that decrease wind speed, yet allow it to go across the inside of the building. This façades also allows the people to see the activities inside the building and merge the public exterior life with the recreational and sportive activities.

立面与材料

设计师采用了一个几何平行条状的屋面系统,以应对和掌控阳光移动路线。这些平行的长条形成技术、空间和生物气候模式,可以向东西方向延伸,也可以在南北方向上增加数量,还可以根据不同体育运动的需要升高,或者为了遮阳而降低。

这些长条形成一个 7 排平行几何金属桁架系统,其改变直接与城市的山区或地形的环境相关。通过三个建筑体量组合形成的金属体量系统,支撑起屋面系统,作为雨水槽将水引入城市排水系统。独特的建设细节、穿孔立面、突出的长条等,设计师通过反复研究解决了生物气候、空间和相关方面的问题。体育场馆不需要太多的机械化配置,利用开孔的立面降低风速,使风在建筑内部贯穿而过。这样的立面设计也使人们既能看到室内的赛事,又能参与到户外公众的娱乐和体育活动中。

KEYWORD 关键词

Skin and Facade 表皮与立面 | Mixed Facade 混合立面

Materials 材料 | Rebar, Glass, Concrete, Aluminum 钢筋、玻璃、混凝土、铝材

Structure 结构 | Reinforced Concrete With Prefab Elements, Steel Roof Structure (Stadium & Sports Hall), Reinforced Concrete Structure (Shopping Centre)
钢筋混凝土预制元素，钢质屋顶结构（体育馆、足球场）
钢筋混凝土结构（购物中心）

Location: Ljubljana, Slovenia
Client: Ljubljana City Municipality, Grep
Architects: SADAR + VUGA
Site Area: 182,600 m²
Photography: David Lotric, Matevz_Lenarcic_Aerovizija, sadar vuga archive

项目地点：斯洛文尼亚卢布尔雅那
客户：卢布尔雅那市政
建筑设计：斯洛文尼亚 SADAR + VUGA 建筑师事务所
占地面积：182,600 m²
摄影：David Lotric, Matevz_Lenarcic_Aerovizija, sadar vuga archive

The Sports Park Stožice
Stožice 体育公园

Features 项目亮点

The football stadium is 'sunken' in the park – only the roof over the stands rises above the plane of the park as a monolithic crater reminiscent of the pit that once formed part of the landscape; A canopy encircles the hall mirrors the scalloped shell.

足球场的表皮采用绿色屋顶与金属材质的混搭，形态如弹坑一般惟妙惟肖；体育馆则采用玻璃幕墙与铝板材质，突出扇形贝壳的衍生结构。

Overview 项目概况

The 187,500sqm Sports Park Stožice is located north of the city and combines a football stadium and a multi-purpose sports hall with a large shopping centre covered by a recreational park landscape. The two storeys of the shopping centre and the indoor car park occupy the 12-metre deep disused gravel pit that once characterized the site. The park changes the periphery of Ljubljana by merging green space and recreational facilities in order to create new areas for social interaction and communal activities.

Stožice 体育公园位于城市北部，占地 187 500 m²，包括一个足球场和一个带大型购物中心的多功能体育馆。双层购物中心和内部停车场刚好设计在一个 12m 深的废弃的巨大砂砾坑内，覆盖在购物中心上部的绿色公园延续着穿越外环路的自然景观，并与市区的其他绿色空间融成一体。

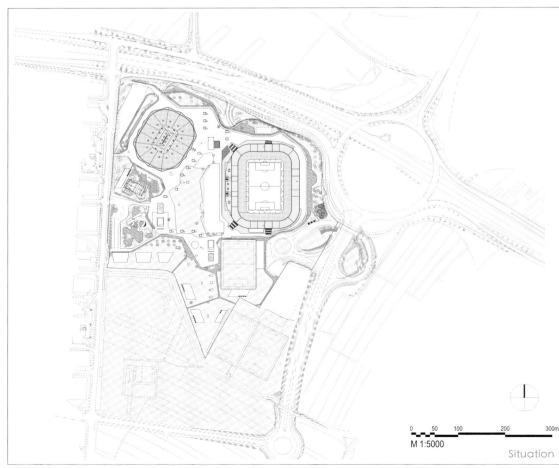

Situation
M 1:5000

Form and Shape   形态与造型

The 16,000 seat football stadium is designed according to contemporary economic, sociological and environmental UEFA standards. Its design and shape provide ideal conditions for sporting events and ensure the perfect experience for the spectators. The football stadium is 'sunken' in the park – only the roof over the stands rises above the plane of the park as a monolithic crater reminiscent of the pit that once formed part of the landscape.

根据目前的经济、社会现状和欧洲足球协会联盟的环境标准，足球场设计为一个可容纳16,000名观众的空间。其设计和外观为体育赛事提供了理想的条件，确保了观众的完美体验。足球场如同公园的一处凹陷区域——只有高于看台的屋顶部分突出地面，从上空俯视，它如同一个巨大的弹坑。

Structure and Materials   结构与材料

The plane is pierced by four flights of stairs that provide access and lead to the stadium's concourse. The park's plateau gradually descends to the east revealing the stadium as an open building, while the eastern entrances create the view into the interior of the stadium. A steel roof stretches high above the stands highlighting its 4x4m grid structure; thus enhancing the quality and sensation of the interior space.

The sports hall for 12,000 spectators is located in the north-western part of the park. The four levels of concourses and the lower, VIP, and upper stands are covered by a shell-shaped dome. This outlines the shape of the hall, a shell that opens towards the perimeter with large crescent openings overlooking the park. A canopy encircles the hall mirrors the scalloped shell. Like the stadium, the entire shell of the hall is also finished in exterior cladding that changes colour depending on the exterior conditions and viewing distance. The sports hall is a partially recessed building whose volume is determined by the required seating capacity and the size of the basketball and handball court.

The position of the stands ensures maximum compactness of the interior space and allows the spectators to be as close to the action on the court as possible. VIP seating is situated between the lower and the upper stands. The warm-up hall is connected to the main hall on the lowest floor, under the surface of the park. The whole interior surface of the shell is determined by the geometry of the steel structure - a large ornament that levitates above the upper concourse, stands, and courts. The lower, VIP, and upper concourses with kiosks surround the interior of the hall and open towards the park. The double height spaces and terraces interconnect visually the concourses and create a permeable space between the interior of the hall and the park.

足球场所在的地面被四个阶梯分割并围合，也形成了体育场的入口处。公园的地势自东往西逐渐下降，所以体育场是一个开放式建筑。从东侧的入口可以看到体育场内部活动，这也进一步突出了体育场是一个位于地面之下的凹陷式建筑的概念。高于看台的钢结构屋顶向上延伸，突出了体育场长4m宽4m的网格结构，从而强化了内部空间的品质及其所带来的体验。

体育馆位于公园的西北角，可容纳12 000名观众，共四层，有个贝壳式的穹顶。穹顶边缘处的波浪状的褶皱形成了正对着公园的开口，褶皱围绕着整个穹顶外围而设计，看起来像是扇形贝壳的衍生结构。与足球场一致，其颜色会随着室外情形和观看距离而不同。体育馆是一个部分凹陷下去的建筑。它的体积由座位数量和比赛场地的大小决定。

看台的位置保证了室内拥有最大的简洁性，并让观众能够更接近地看到比赛。VIP座位设置在下层看台和上层看台之间。热身场馆与主厅最底层连接。足球馆室内表层是根据钢铁结构的几何形状设计的———一个位于整个体育馆上空的巨型装饰物。下层区、VIP区和上层区都设有能看到公园的开口，这种设计让整个建筑具有一种渗透性，使内外空间融通相连。

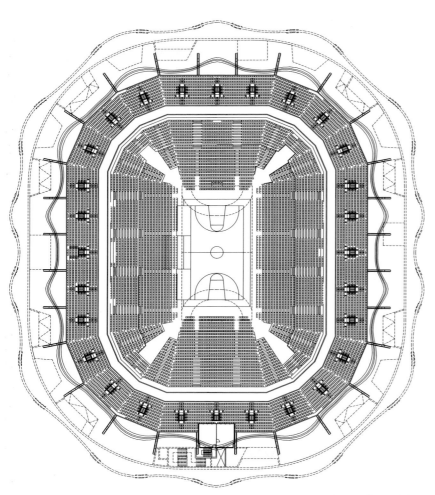

Arena Plan +14

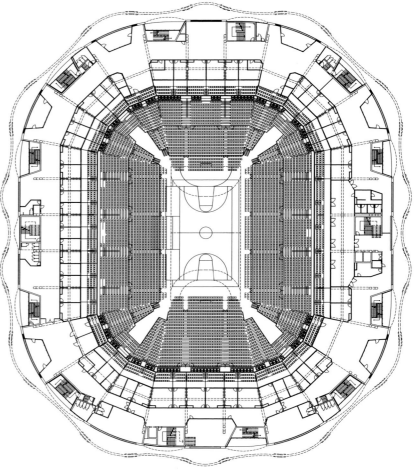

Arena Plan +5

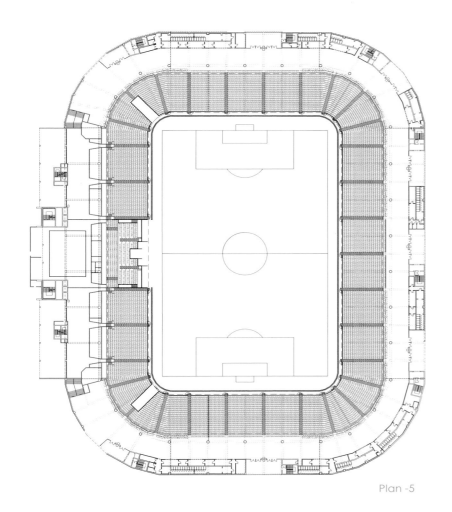

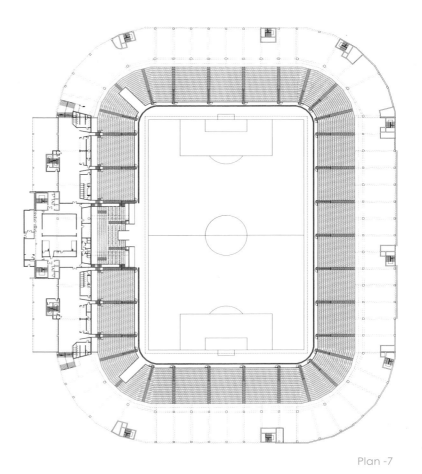

Plan -5

Plan -7

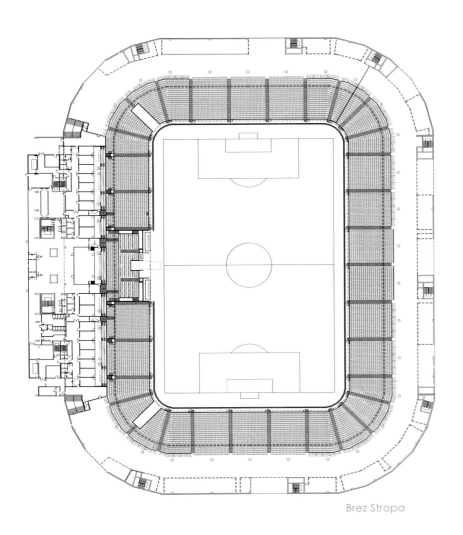

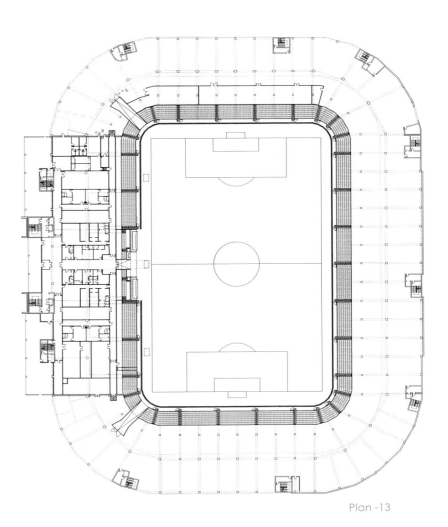

Brez Stropa

Plan -13

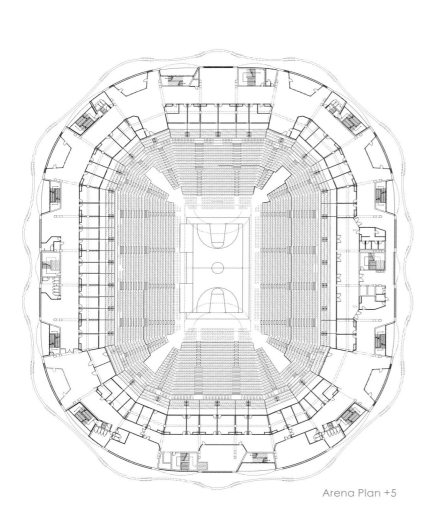

Arena Plan +5

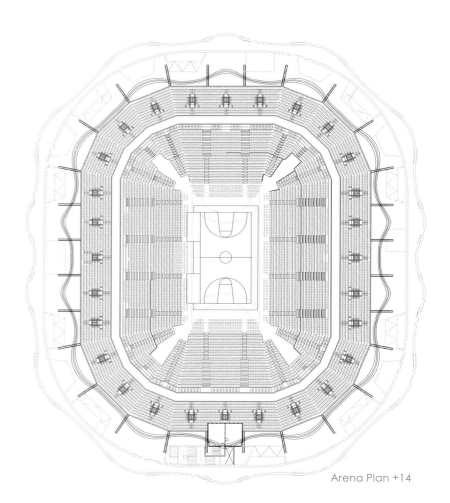

Arena Plan +14

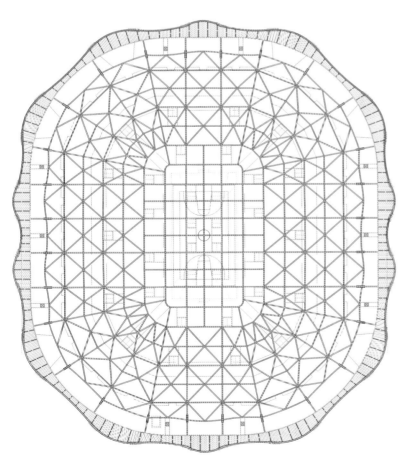

Arena Plan Ostresje

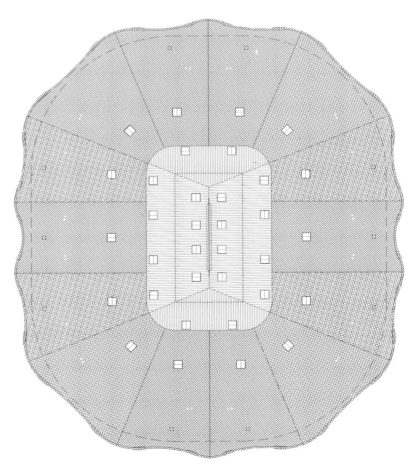

Arena Plan Streha 1

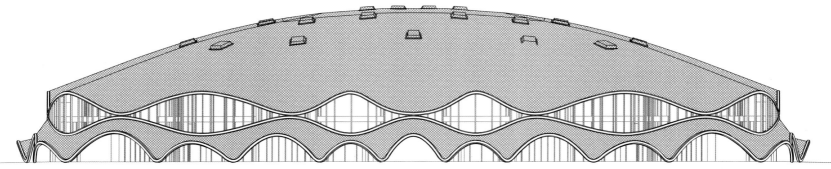

Arena Elevation

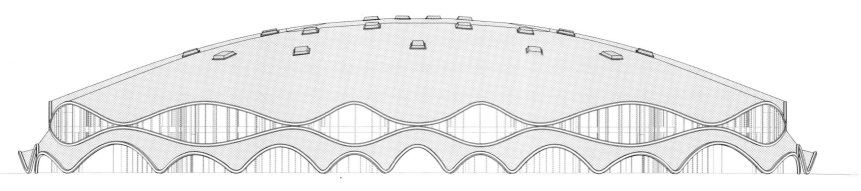

Arena Elevation

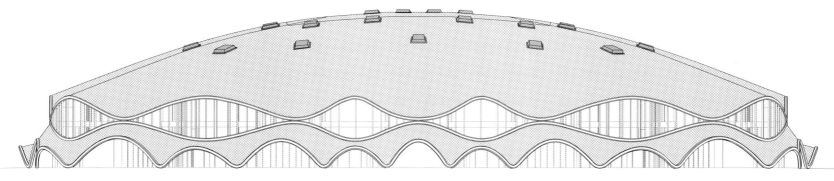

Arena Elevation

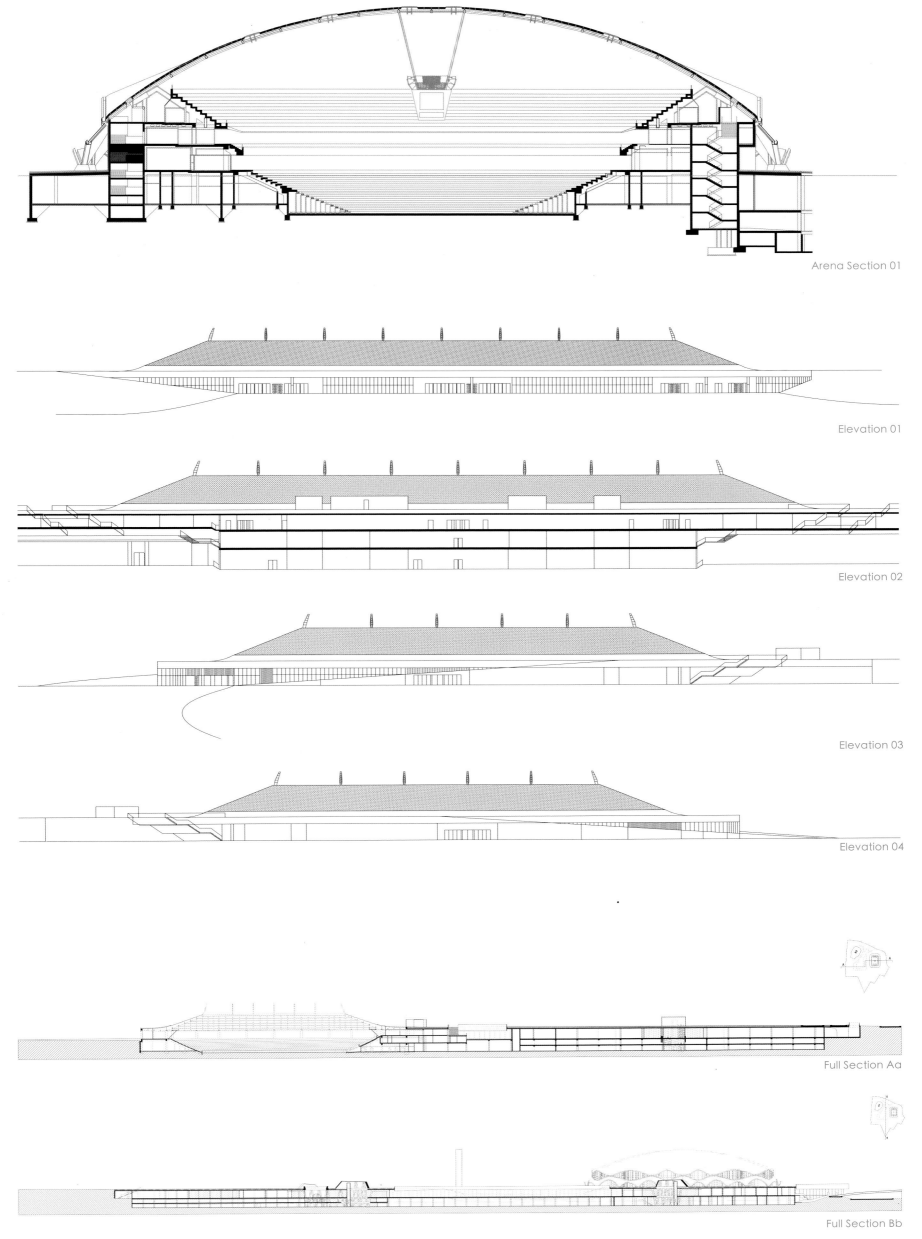

Arena Section 01

Elevation 01

Elevation 02

Elevation 03

Elevation 04

Full Section Aa

Full Section Bb

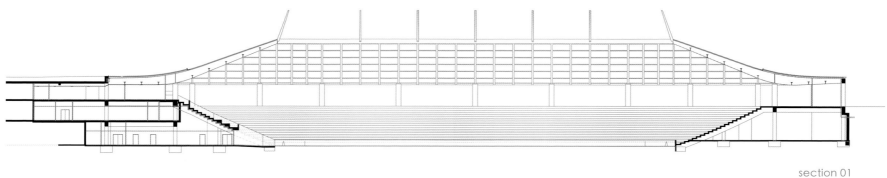
section 01

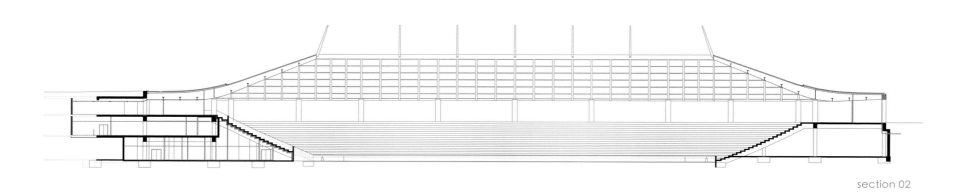
section 02

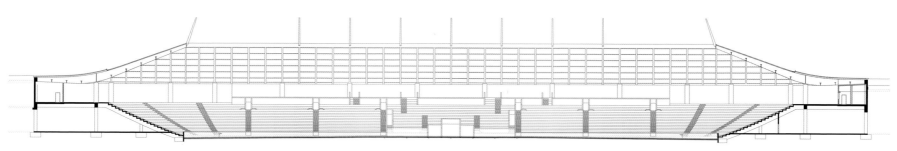
section 04

KEYWORD 关键词	Skin and Facade 表皮与立面	Metal Facade 金属立面
	Materials 材料	Nonferrous Metals, Glass, Steel 有色金属、玻璃、钢
	Detail 细部节点	

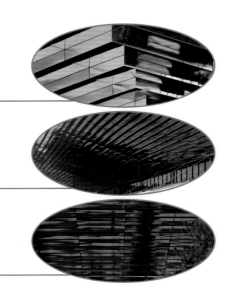

Location: Rouen, France
Client: La Crea, France
Architect: Dominique Perrault Architecture, Paris, France
Site Area: 31,500 m²
Planted Area: 1,600 m²
Photography: Axel Dahl, DPA, Final Cad, Georges Fessy, Kindarena

项目地点：法国鲁昂
客户：La Crea, France
建筑设计：法国巴黎 Dominique Perrault 建筑师事务所
占地面积：31 500 m²
绿化面积：1 600 m²
摄影：Axel Dahl, DPA, Final Cad, Georges Fessy, Kindarena

Sports Palace
皇家体育场

Features 项目亮点

The design emphasizes the modern sense of the contours of the stadium. Outward metal façade offers a strong visual impact. The base of the steps is novel and unique.

设计强调体育场的现代轮廓感，金属表皮的外立面通过不断向外延伸的体量，形成了具有视觉冲击力的外部形态，底座台阶的设计新颖而独特。

Overview 项目概况

The functions are very clearly distinguished. Sports areas are situated in the central part of the building, the reception space is in the south wing of the building along the Lillebonne Street and the administrative and services areas in the north wing. There are two sports halls with flexible seating capacity for up to 6 000 and 864 people respectively.

这是由巴黎 Dominique Perrault 建筑师事务所设计的法国鲁昂皇家体育场，共有两个体育场馆。体育场的功能区域划分明确，运动区位于中央部分，接待空间在该建筑的南部边缘，行政服务区在北部边缘。两个场馆的座位布置灵活，分别可容纳 6 000 人和 864 人。

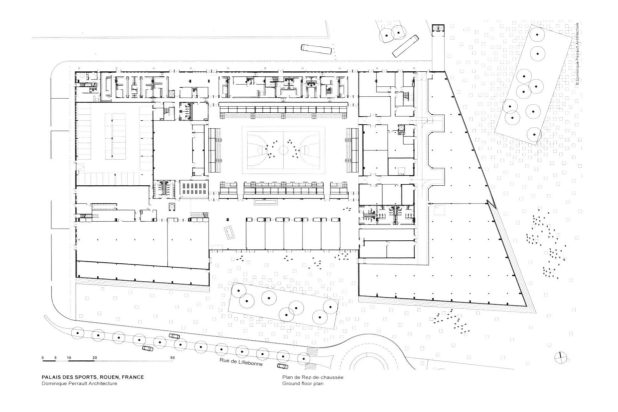

PALAIS DES SPORTS, ROUEN, FRANCE
Dominique Perrault Architecture

Plan de Rez-de-chaussée
Ground floor plan

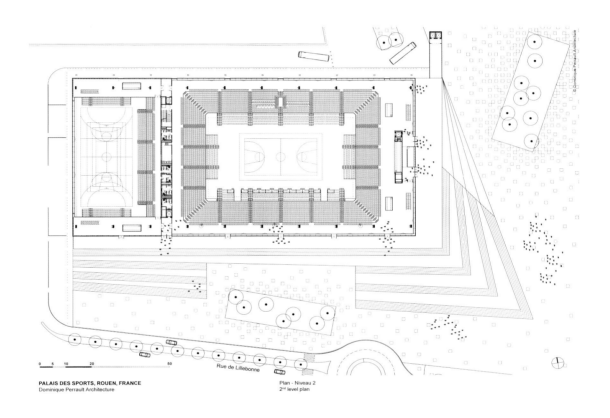

PALAIS DES SPORTS, ROUEN, FRANCE
Dominique Perrault Architecture

Plan - Niveau 2
2nd level plan

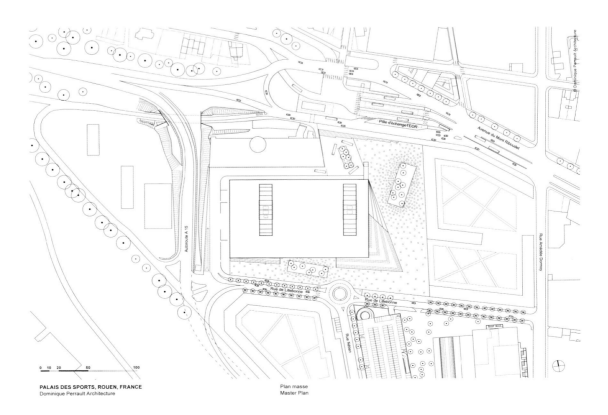

PALAIS DES SPORTS, ROUEN, FRANCE
Dominique Perrault Architecture

Plan masse
Master Plan

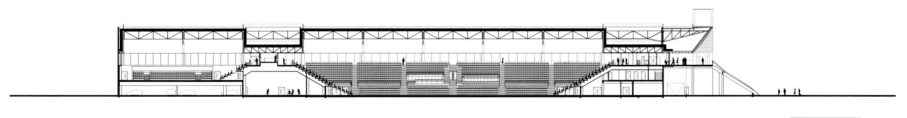

PALAIS DES SPORTS, ROUEN, FRANCE
Dominique Perrault Architecture

Coupe longitudinale
Longitudinal section

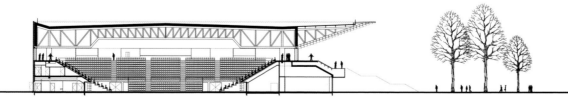

PALAIS DES SPORTS, ROUEN, FRANCE
Dominique Perrault Architecture

Coupe transversale
Cross section

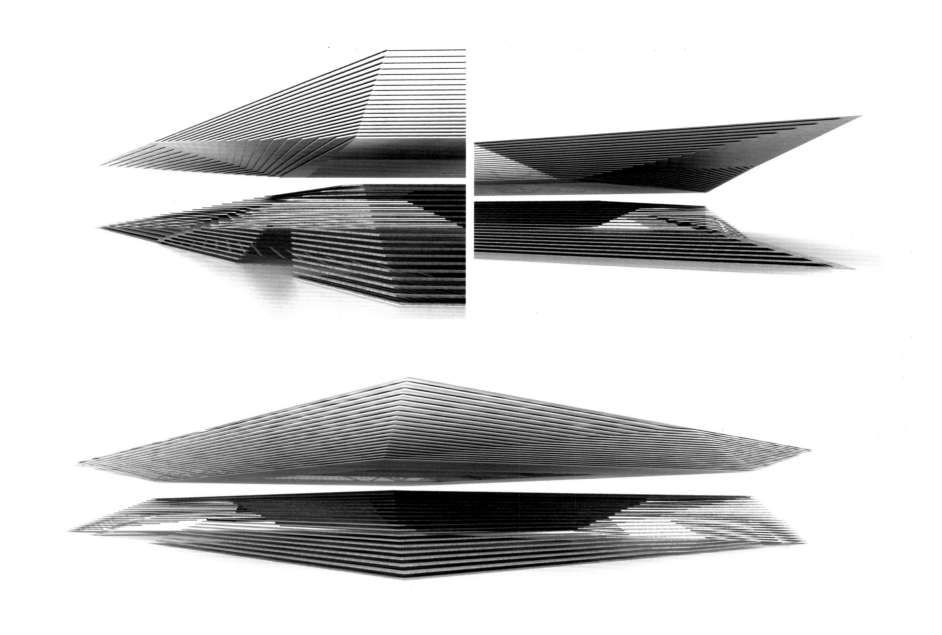

Function and Layout 功能与布局

- main sports hall : 4,400 m²

audience capacity 5,650 (for Basket ball configuration)

maximum audience capacity 6,000 (for Boxing configuration)

- sports annex : 2,400 m²

secondary gym space, maximum audience capacity 864

- audience annex : 2,200 m²

ticket office, shops, refreshment area, infirmary, etc.

- annex to sporting facilities : 1,300 m²

meeting rooms, changing rooms, antidoping room, infirmary and massage room, offices for trainers and referees

- reception and media space : 2,800 m²
- services : 470 m²
- technical facilities : 1,400 m²

maintainance and stock rooms

- car park : 1,000 m² / 41 vehicles

主体育馆面积约4 400 m²，篮球场馆可容纳观众5 650人，拳击场馆可容纳观众6 000人。运动附加场所包括二级健身房约2 400 m²，最多可容纳864名观众，其售票处、商店、茶点区、医务室等约占2 200 m²，会议室、更衣室、体检室、按摩室、裁判和培训人员办公室等约1 300 m²。停车场面积为1 000 m²，约41个车位。

KEYWORD 关键词	Skin and Facade 表皮与立面	Glass Facade 玻璃立面
	Materials 材料	Curtain Wall / Aluminum Composite Panel 合成玻璃、铝复合板
	Structure 结构	Reinforced Concrete 钢筋混凝土

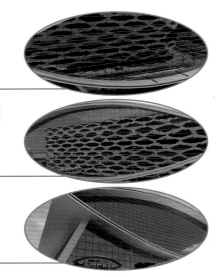

Location: Manila, Philippines
Clientr: SM Investment Corporation
Design Architect: Arquitectonica
Total Site Area: 172,158 SF / 15,994 SM
Total Area (GFA): 561,876 SF / 52,216 SM
Photographer(s): Pathfinder

项目地点：菲律宾马尼拉
客户：SM 投资公司
建筑设计：Arquitectonica
总用地面积：172 158 SF /15994 M²
总面积（GFA）：561 876 SF /52 216 M²
摄影师：Pathfinder

Mall of Asia Arena
亚洲商城体育馆

Features 项目亮点

With semi-elliptical shape structure, aluminum composite panel material for the top and glass facade, the building has a functionally-purposed three-dimensional layout.

外形上呈半椭圆形的建筑结构，顶部表皮采用铝复合板，立面运用玻璃作为主要的材料，整体采用立体式的布局，功能明确。

Overview 项目概况

The 52,200 sqm Mall of Asia Arena is a 15,000 seat multi-purpose venue located in Manila. It was designed to cater primarily for concerts and basketball games, but can also be configured to allow for boxing, theater, fashion, and ice-skating shows.

亚洲商城体育馆是一个功能型的场馆。位于菲律宾马尼拉，总面积为52 200 m²，场馆共有15 000个座位。场馆设计主要用于音乐会和篮球比赛，适当调整后也可以作为拳击、戏剧、时装、溜冰场所。

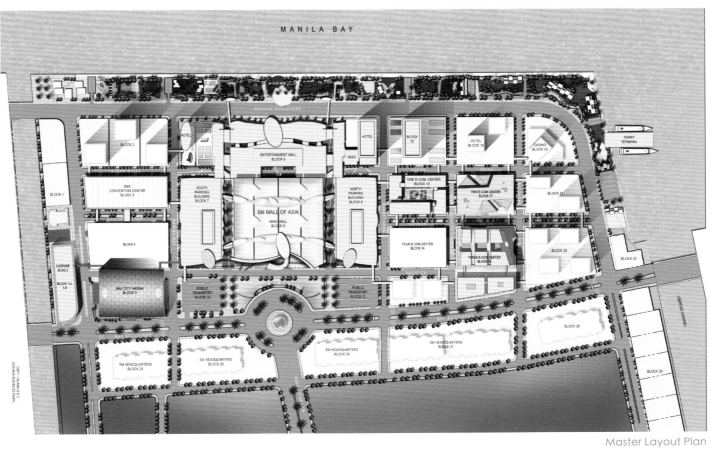

Master Layout Plan

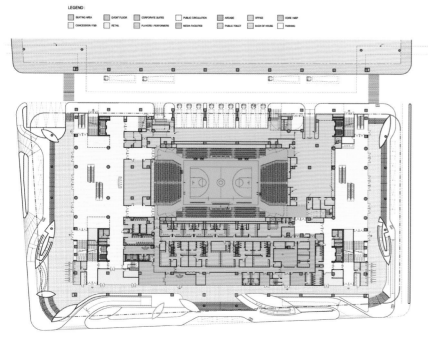

L1 Floor Plan

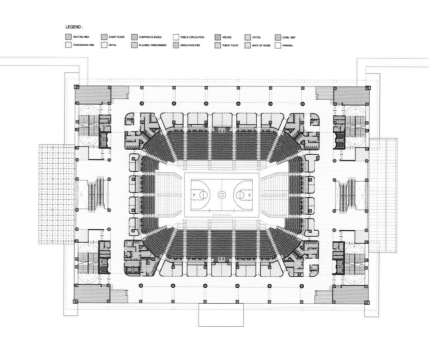

L2 Floor Plan

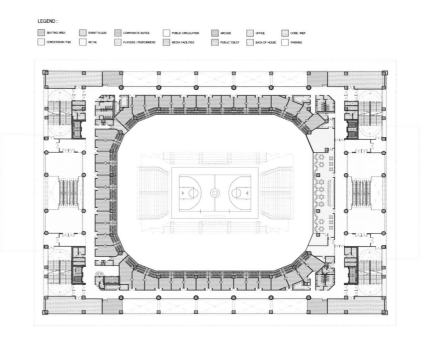

L3 Floor Plan

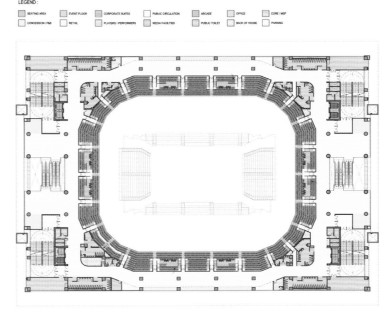

L4 Floor Plan

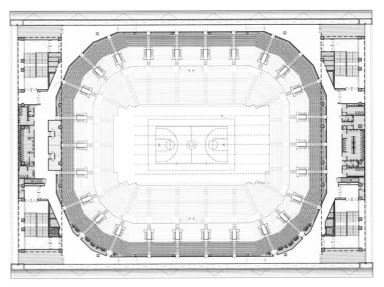

L5 Floor Plan

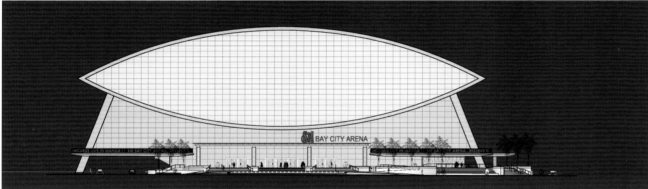

Mall of Asia Arena_Elevation South

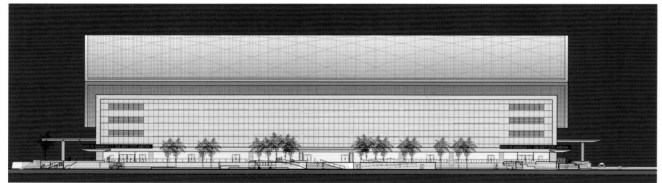

Mall of Asia Arena_Elevation East

Structure and Layout　　结构与布局

Arranged on two tiers, it has a ring of corporate suites separating the lower seats from the balcony levels. Backstage, changing, and press areas are located on the ground floor, while BOH and admin areas are located on interstitial mezzanines. Major concourses are lined with concessions, while a VIP restaurant and lounge is located on the corporate suite level.

体育馆分为两层，位于中间的环形工作区将上下两个坐席区分开。后台、更衣室和媒体发布室位于一楼，场馆工作区和管理区位于中间夹层。两个主通道位于夹层上下，VIP 餐厅和酒廊位于工作区。

LEGEND:

- SEATING AREA
- EVENT FLOOR
- CORPORATE SUITES
- PUBLIC CIRCULATION
- ARCADE
- OFFICE
- CORE / MEP
- CONCESSION / F&B
- RETAIL
- PLAYERS / PERFORMERS
- MEDIA FACILITIES
- PUBLIC TOILET
- BACK OF HOUSE
- PARKING

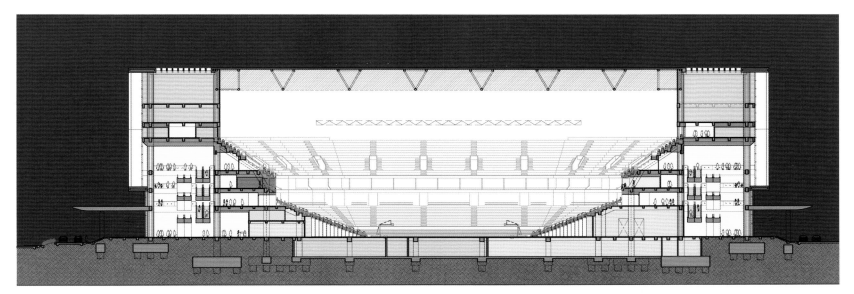

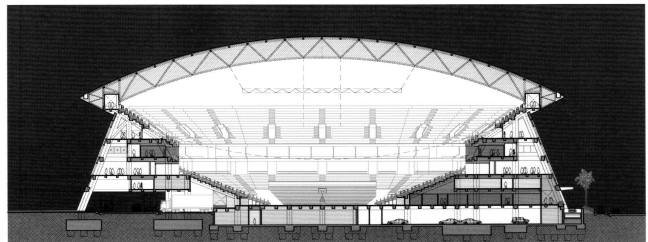

Mall of Asia Arena_Section

KEYWORD 关键词	Skin and Facade 表皮与立面	Mixed Facade 混合立面
	Materials 材料	Wood, Glass, Steel 木材、玻璃、钢材
	Shape 形态	Curved Roofs 曲面屋顶

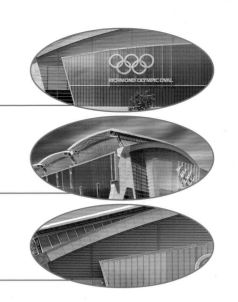

Location: Vancouver Canada
Architectural Design: Cannon Design
项目地点：加拿大温哥华
建筑设计：坤龙建筑设计公司

Richmond Olympic Oval
列治文奥运速度滑冰馆

Features 项目亮点

Using wood, glass and steel, the build has a steel and glass mixed outlook and a curved roof.

本案在设计时将顶部设置为曲面的屋顶，整体结构上运用了木材、玻璃和钢材，外立面突出钢与玻璃的混搭装饰。

Overview 项目概况

Richmond Olympic Oval designed for the 2010 Vancouver Winter Olympics by Cannon Design, covers an area of 512,000 square feet and 8,000 seats, a 400-meter speed skating rail, VIP lounge and anti-doping laboratory. An important design goal: permanent.

列治文奥运速度滑冰馆（Richmond Olympic Oval），由美国坤龙建筑设计公司（Cannon Design）为2010年温哥华冬奥会设计，占地面积47 566 m^2，8 000个座位。一条400m的速滑道，还有贵宾休息厅和反兴奋剂实验室。永久性是该馆设计的一个重要的目标。

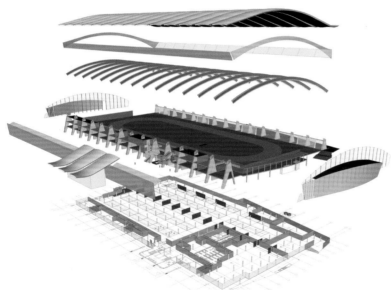

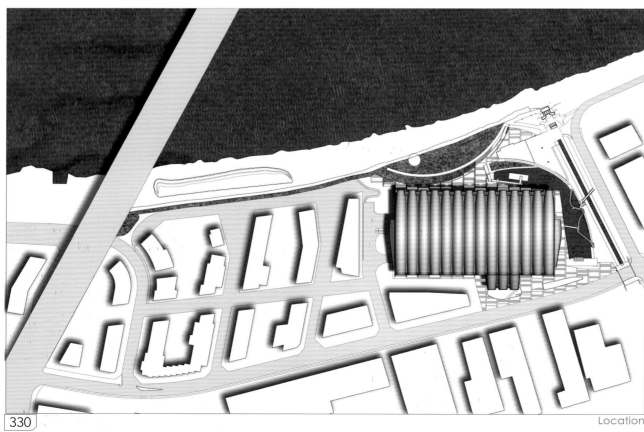

Location

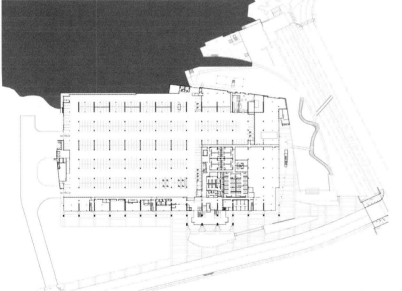

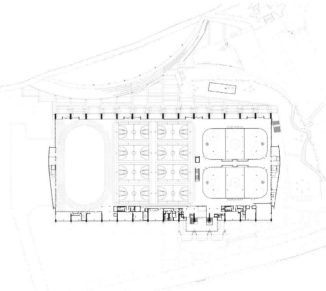

Plan-Level1

Plan-Level2-Legacy

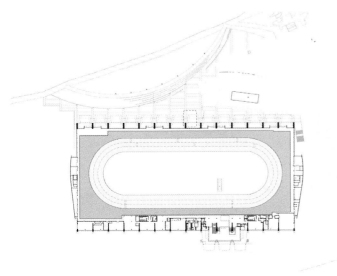

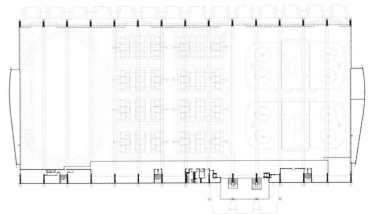

Plan-Level2games

Plan-Level3

Shape and Materials
造型与材料

Using wood, glass and steel in combination, its most prominent feature is the 6.5 acres of the curved roof, which uses insecticide-treated pine from British Columbia. It is the first time pest timber been used.

Cannon Design hopes this $ 63.3 million building won LEED-Silver Award. After the Olympics, it will be used as a community center and training center for athletes.

这幢建筑将木材、玻璃和钢材结合起来使用。它的最突出的特色，是面积约为 26 304 m² 的曲面屋顶，使用了来自不列颠哥伦比亚省的经过杀虫剂处理的松木。这是首次使用有过虫害的木材。

希望这幢 6 330 万美元的建筑物赢得绿色建筑银奖（LEED-Silver）认证。奥运会之后，它将作为社区活动中心和运动员训练场地。

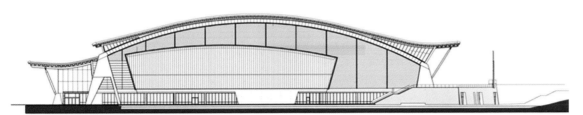

East-Elevation

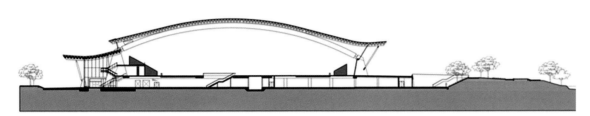

Section

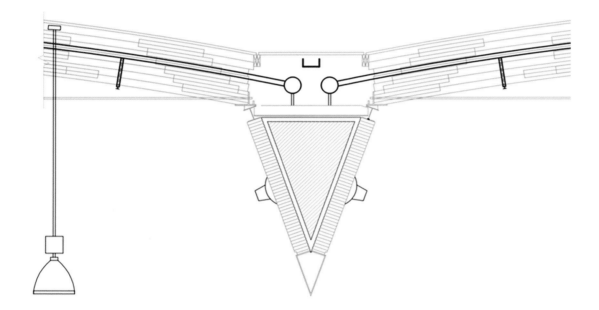

Roof-Detail

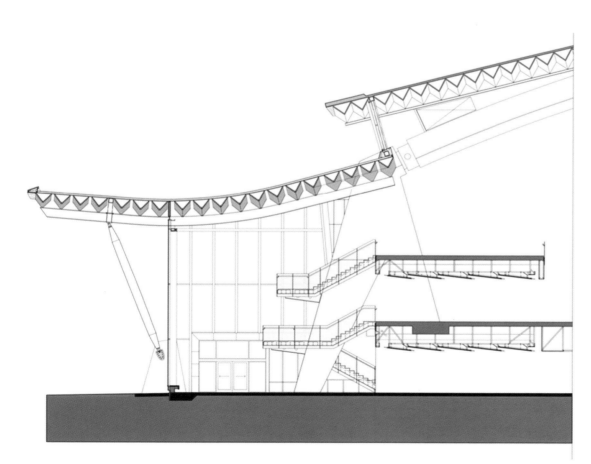

Lobbysection

KEYWORD 关键词		
Skin and Facade 表皮与立面	Metal Facade 金属立面	
Materials 材料	Aluminum, Steel 铝材、钢	
Shape 造型	Magic Box "魔盒"造型	

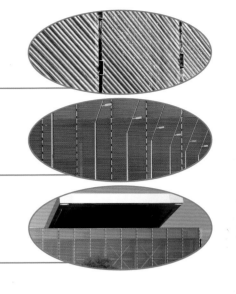

Location: Madrid, Spain
Client: Madrid Espacios y Congresos
Architect: Dominique Perrault Architecture
Built Area: 100,000 m²
Photography: Georges-Fessy, DPA

项目地点：西班牙马德里
客户：Madrid Espacios y Congresos
建筑设计：法国 Dominique Perrault 建筑师事务所
建成面积：100 000 m²
摄影：Georges-Fessy, DPA

Olympic Tennis Centre, Madrid, Spain
西班牙马德里奥林匹克网球中心

Features 项目亮点

Inside the "magic box" the tennis arenas are adapted to the different uses of the complex. The roofs of the three indoor/outdoor courts are giant mobile slabs mounted on hydraulic jacks, which interplays with the light skin.

建筑总体呈盒子形状，设计时将顶部设置为开放式的结构以满足多种用途，上部巨型移动平板与立面可移动的轻质表皮相得益彰。

Overview 项目概况

The Olympic Tennis Center is located in a former slum housing area in the middle of a busy motorway and train network. The project aims to reinforce the Spanish Capital's candidature for the 2016 Olympics.

The built project includes the "magic box" with three indoor/outdoor courts, with covered area for 20,000 spectators, 16 outdoor courts, five courts with a covered area for 350 spectators each, six practice courts, a pool, headquarters for the Madrid Tennis Federation, a tennis school, clubhouse, press center, stadium boxes and other private areas and restaurants.

　　该奥林匹克网球中心所在地原是一个贫民区，周围是高速公路和列车轨道交织的交通网络。项目旨在助西班牙首都申办2016年奥运会一臂之力。

　　项目包括拥有三个室内外网球场可容纳20 000名观众的"魔盒"、16个室外网球场、5个分别可容纳350名观众的有盖球场、6个练习球场、一个泳池、马德里网球联合会、一所网球学校、会所、新闻中心、体育场、其他私人场所和餐厅。

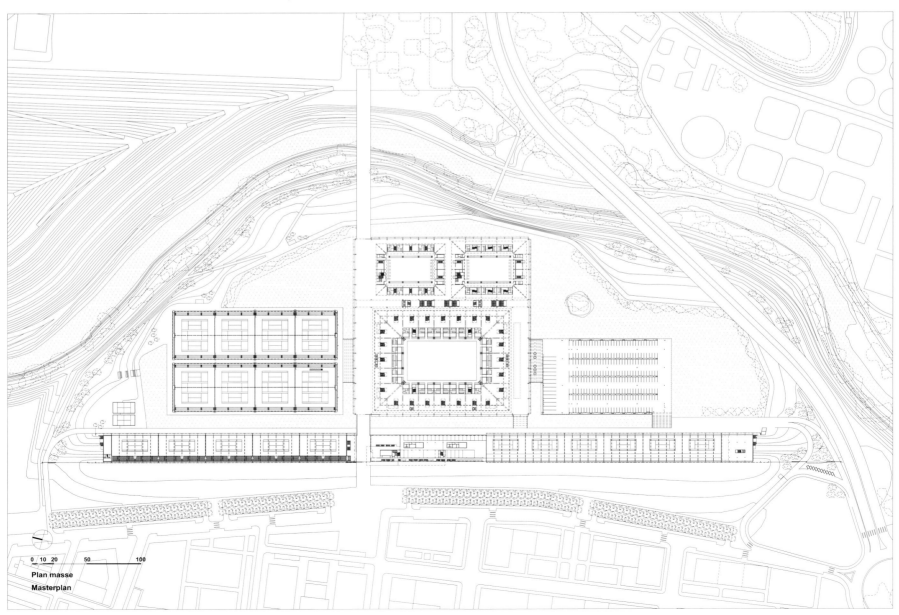

Plan masse
Masterplan

Madrid Plan Masse

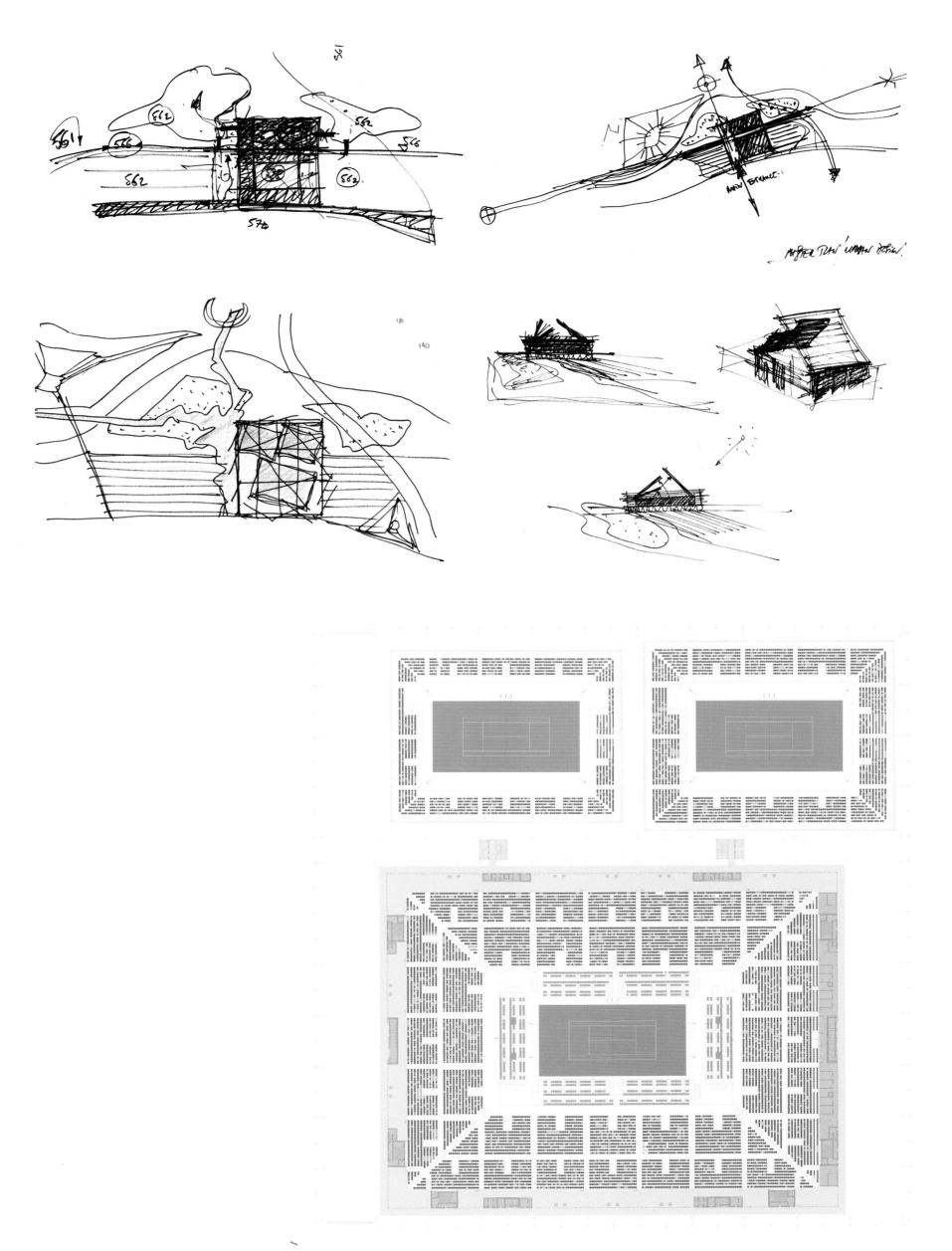

COT_PLAN_LEVEL +3

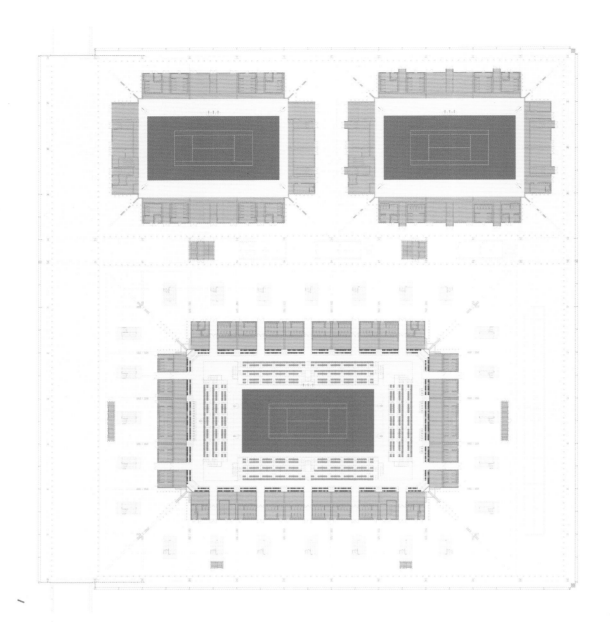

COT_PLAN_LEVEL-0

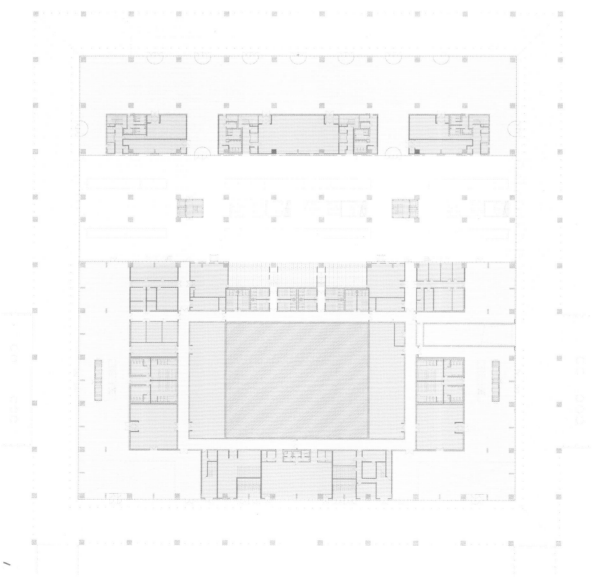

COT_PLAN_LEVEL -2

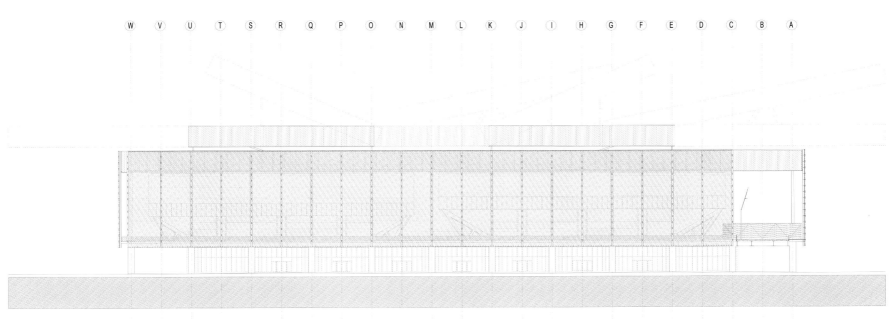

COT_ELEVAT_E

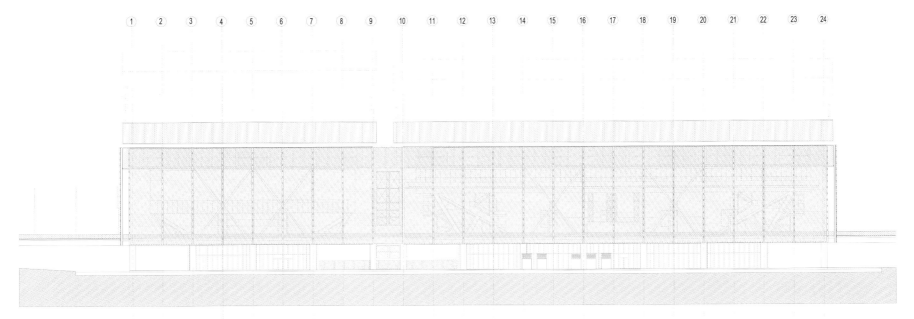

COT_ELEVAT_N

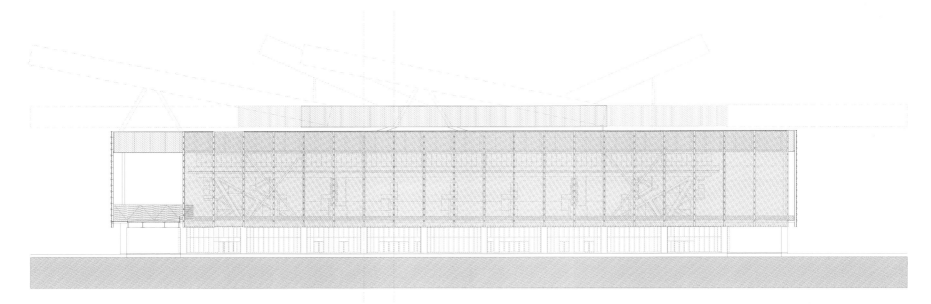

COT_ELEVAT_W

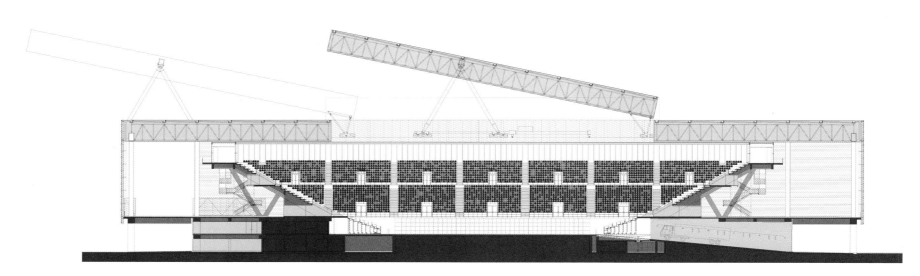

COT_SECT_15

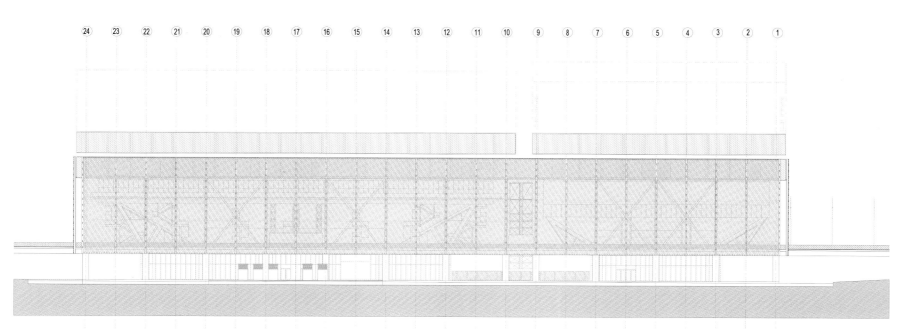

COT_ELEVAT_S

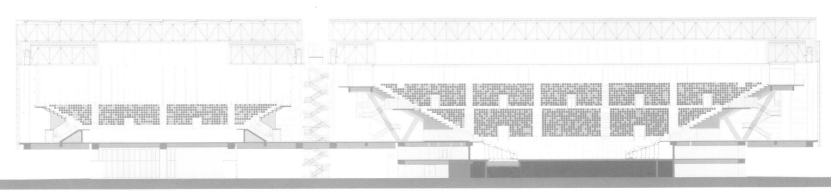

COT_SECT_P

Shape and Skin 造型与表皮

The "magic box" concept encloses sports and multi-functional buildings but opens up and shapes itself to the various uses projecting a changing and lively silhouette in the cityscape. Its mobile and vibrant skin filters the sunlight, serves as a windbreak and shelters the sports halls in a lightweight shell.

"魔盒"概念围绕运动和多功能建筑展开,同时将自身融入多种用途,活力四射,在城市景观中脱颖而出。可移动、充满活力的表皮不仅能过滤灼热的阳光,还充当着防风墙的角色,保护场馆。

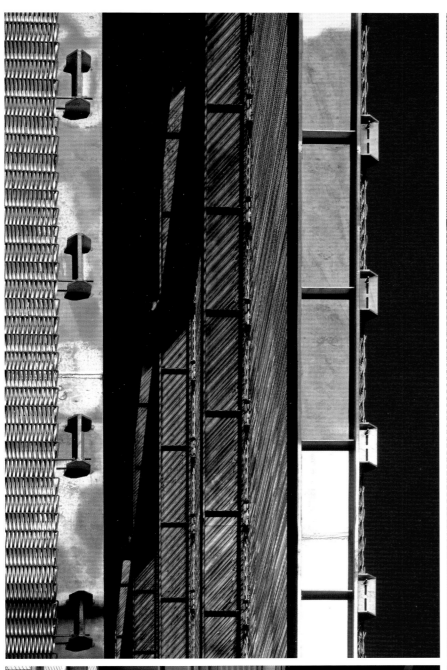

Structure and Function
结构与功能

Inside the "magic box" the tennis arenas are adapted to the different uses of the complex. The roofs of the three indoor/outdoor courts are giant mobile slabs mounted on hydraulic jacks, which serve to partially or totally open the three roofs to allow for passage of air and sunlight, or close them to avoid exposure to the rain or other hazardous weather conditions.

Even in the worst weather conditions, Madrid's Olympic Tennis Center can hold a minimum of three simultaneous matches. This versatility allows it not only to celebrate almost any kind of sports meeting, but also a significant number of other events, such as concerts, political meetings, fashion shows, etc.

"魔盒"内部的网球场地能适应不同的用途。三个室内外网球场的顶部是由液压千斤顶固定的巨型移动平板，晴时打开雨时闭。

即使在最坏的天气情况下，该网球中心都能同时举行三场比赛。这种多功能性使得它不仅能举行各种体育赛事，还能展开一些其他重要活动，如音乐会、政治会议和时装秀等等。

KEYWORD 关键词		
Skin and Facade 表皮与立面	Mixed Facade 混合立面	
Materials 材料	Glass, Steel 玻璃、钢材	
Shape 造型	Waved Shape 波浪造型	

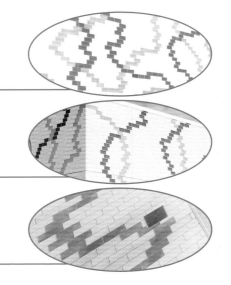

Location: Montreal, Quebec, Canada
Architect: Saïa Barbarese Topouzanov Architectes
Photography: Marc Cramer, Frédéric Saïa, Vladimir Topouzanov

项目地点：加拿大魁北克蒙特利尔
建筑设计：Saïa Barbarese Topouzanov 建筑设计事务所
摄影：Marc Cramer, Frédéric Saïa, Vladimir Topouzanov

Expansion of Centre Sportif J.C. Malépart

J.C. Malépart 体育中心扩建项目

Features 项目亮点

Both indoors and outdoors, a wave develops on the perimeter of the building, rising and falling in two continuous undulations. Glass façade, milky tone and reflective quality of the upper wall are echoed in the material for the roof to ensure the extension from one to the other in contrast with the clear demarcations of the building erected fourteen years ago.

在造型设计上，室内外均采用波浪形态，玻璃立面、不透明牛奶色的墙体与屋顶材料色调相呼应，建筑整体简洁现代而具有可辨性。

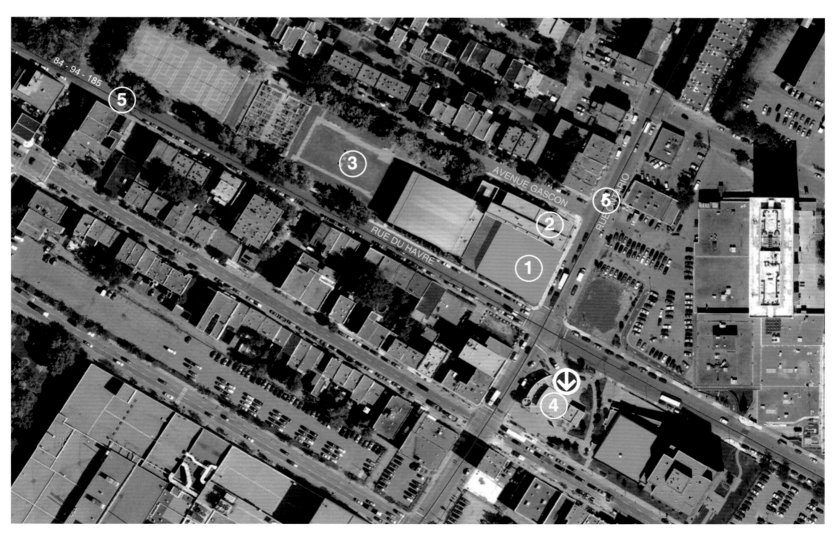

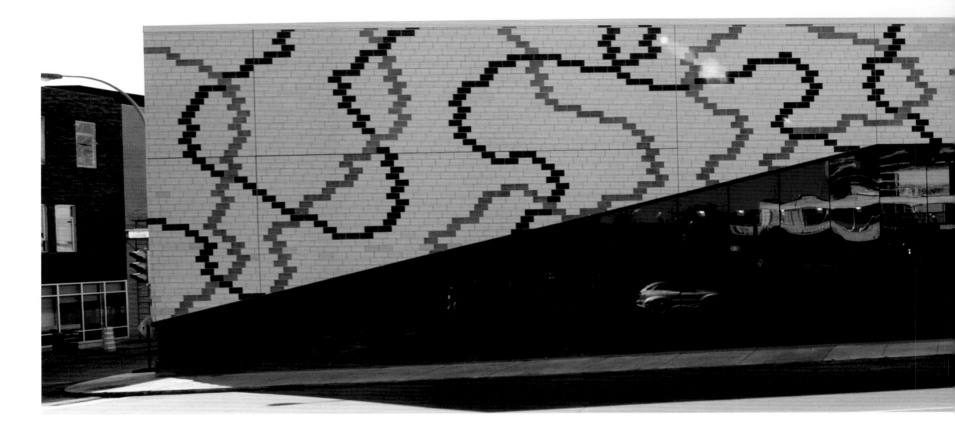

PLAN D'IMPLANTATION

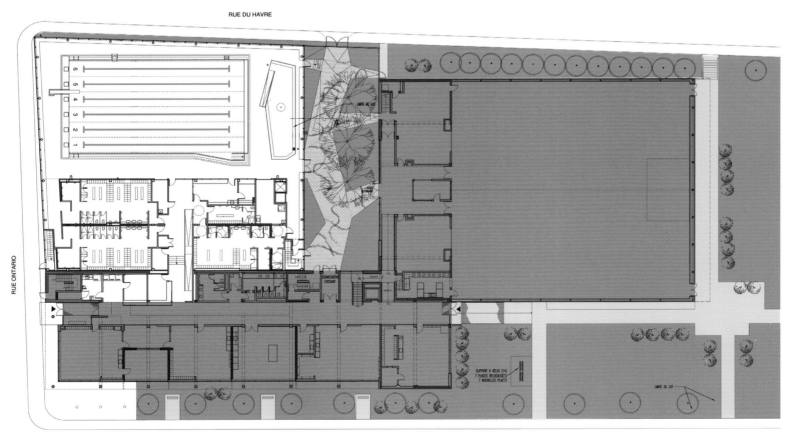

ÉCHELLE 1:300

PLAN DU REZ-DE-CHAUSSÉE

LÉGENDE
A ACCUEIL
B VESTIAIRES FEMMES
C DOUCHES FEMMES
D VESTIAIRES HOMMES
E DOUCHES HOMMES
F VESTIAIRES FAMILLES
G DOUCHES
H DOUCHES HANDICAPÉS
I VESTIAIRES EMPLOYÉS
J INFIRMERIE- RÉGIE
K BUREAU
L ENTREPOSAGE
M PATAUGEOIRE
N BASSIN NATATION

ÉCHELLE 1:200

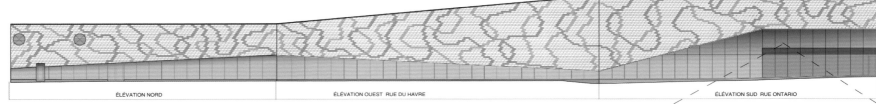

| ÉLÉVATION NORD | ÉLÉVATION OUEST RUE DU HAVRE | ÉLÉVATION SUD RUE ONTARIO |

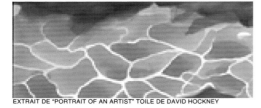

EXTRAIT DE "PORTRAIT OF AN ARTIST" TOILE DE DAVID HOCKNEY

À l'intérieur comme à l'extérieur, une vague se développe sur le périmètre de l'édifice. Ses mouvements ascendants et descendants se suivent en deux ondulations continues. Le motif de vaguelettes et de reflets d'eau trouvent leur écho sur les plages de la piscine et de la pataugeoire.

VUE DU PONT DEPUIS LA SALLE D'ENTRAINEMENT

Une vague se développe sur le périmètre de l'édifice. Sa forme rappelle la silhouette du pont Jacques-Cartier visible de la piscine. Ses mouvements ascendants et descendants se suivent en un ruban continu.

COUPES

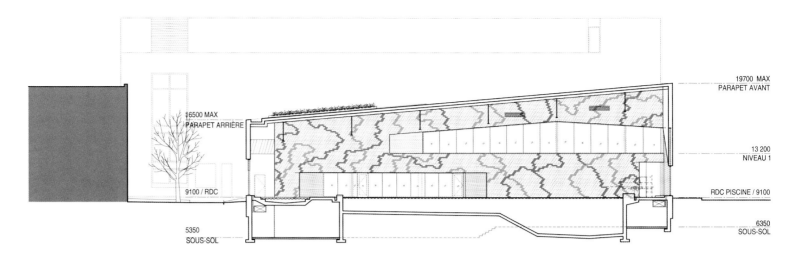

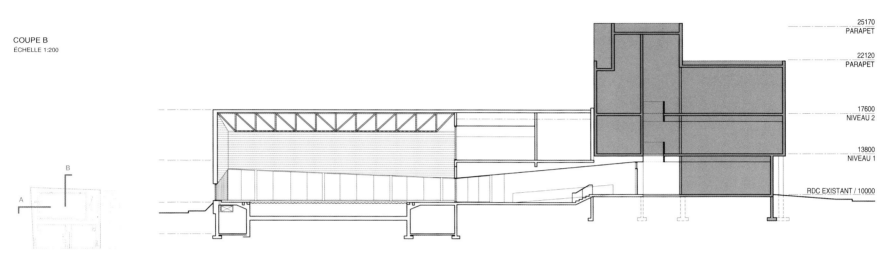

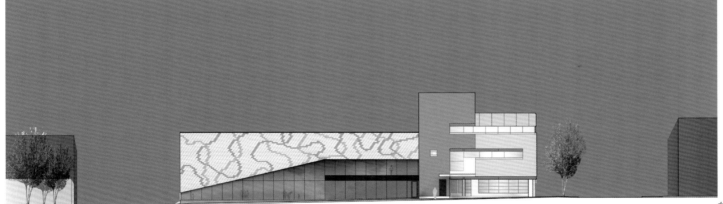

ÉLÉVATION SUD - RUE ONTARIO
ÉCHELLE 1:300

ÉLÉVATION OUEST - RUE DU HAVRE
ÉCHELLE 1:300

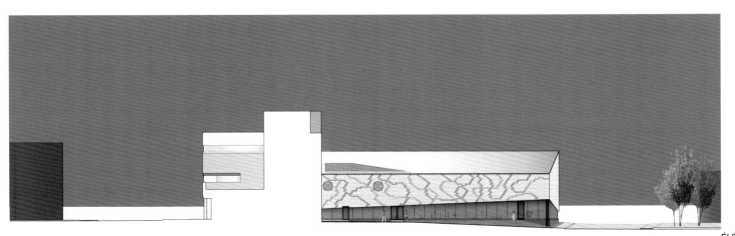

ÉLÉVATION NORD - JARDIN
ÉCHELLE 1:300

Skin and Structure 表皮与结构

The expansion forms a simple, diaphanous volume. It allows clear identification of the entities created and the dialogue between them. On the outside, it adopts the scale of neighboring structures. It reaches its maximum elevation along Rue Ontario. From there, a variable geometry gently inclines the roof to make it compatible with the two- and three-storey structures on Rue du Havre; to refer to the curved roof of the gym; and to provide the gym block with natural light and the adjacent garden with maximum sunlight. This linear, functional garden has become a structuring axis for the various components of the site. The opaque upper part of the skin is in a pale tone that helps to reduce heat loss into the environment.

该扩建项目形成简单而又透明的体量。新建筑与当地建筑群形成鲜明的对比却又彼此和谐地存在着。建筑外部与周围建筑结构类似，延伸到安大略省街道的最远处。建筑侧翼逐渐地往屋顶倾斜以配合当地两层或三层高的建筑结构，同时体育馆的屋顶呈拱形使室内光线良好同时确保了相邻的花园获得充足的阳光。这一线形的多功能花园作为建筑各个功能分区的构建轴线，不透明的建筑表皮呈灰白色调可减少热损失。

Shape and Materials 造型与材料

The idea of waves, inspired by a David Hockney painting, Portrait of an Artist: Both indoors and outdoors, a wave develops on the perimeter of the building, rising and falling in two continuous undulations. The first, transparent and high opposite to the entrance and the training room, surges as far as the corner near the diving board, providing an ambience propitious to concentration. The dialogue between the new building and the old one, continues in the properties of the materials selected. The grey-tinted glass harmonizes with the colours of the earlier structure. It gives passers-by an idea of what is going on inside while preserving some privacy for users. The milky tone and reflective quality of the upper wall are echoed in the material for the roof to ensure the extension from one to the other in contrast with the clear demarcations of the building erected fourteen years ago.

波浪造型的灵感来自大卫·霍克尼的油画——艺术家的肖像。不论建筑外部还是内部都采用波浪造型，如同两道大浪上下起伏。新旧建筑之间的融合体现在建筑材料的选择上。茶色隔热玻璃与当地旧建筑颜色一致，同时确保用户的隐私。牛奶色的墙体与屋顶材料的颜色呼应，同时与旧建筑形成对比，突出扩建项目的可辨别性。

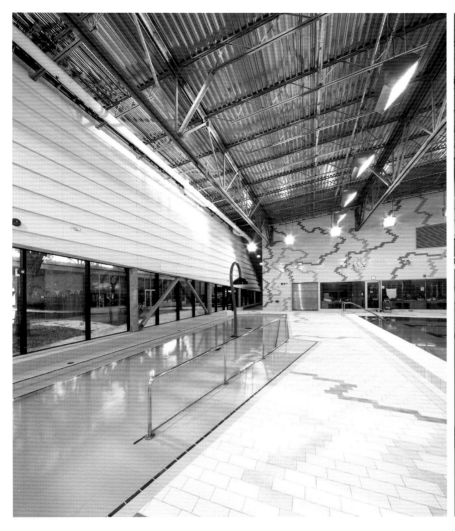

Publicity
3D Structure
Functions
Multiple Layout

公共属性
立体结构
功能突出
多元布局

Healthcare Building
医疗建筑

As a special place for the sick, the weak and the disabled, healthcare building is the result of the medical and technological development. It's an integrated facility to organize and distribute public resources. The same with other new-type buildings, healthcare buildings have ever been the center of cities with their huge volume and solemn style. Today, new healthcare buildings pay much attention to the physical and psychological factors, trying to create a family-like and garden-style environment for the patients.

Architectural design of this kind of building emphasizes the integration of functions and form. Glass and aluminum materials are usually used for facade, shaping an elegant and modern appearance. While the interior spaces should be well organized and optimized to create a peaceful, elegant and human-caring environment for the patients. Healthcare buildings usually apply decentralized, centralized or mixed layout, thus the circulation is an important part of the design. With the sustainable development of the medical science and concept, healthcare buildings will be developed to "large and comprehensive" healthcare center or "small and specific" specialized hospital and clinic.

医疗建筑是现代社会照护病弱伤残成员的特殊场所，它是集中组织、分配公共资源的机构，是医学和技术进步的产品。与新型的建筑物一样，医院建筑也曾经占据城市中心的重要位置，体量高大、风格庄重。如今，由于建立在生理、心理、社会互动的医院模式渐起，以患者为中心的家庭化、园林化的现代医疗建筑引起了社会的关注。

医疗建筑在设计上注重功能与形态的结合，通常采用玻璃和铝材等作为主要的立面材料，外部形态简洁而现代。在内部空间结构上，强调空间的组合与优化，合理的划分了医务人员与患者的空间，为患者创建了幽雅宁静、富有人情味的医疗环境。在布局方面，医疗建筑一般采用分散式、集中式以及混合式的布局形式，交通流线组织往往成为了设计的重要考量部分。随着医疗事业的不断发展和医疗观念的革新，医疗建筑正向着"大而全"综合医院、医疗中心，"小而精"的专科医院、诊所方向发展。

KEYWORD 关键词	Skin and Facade 表皮与立面	Glass Facade 玻璃立面
	Materials 材料	Aluminum, Steel, Glass 铝材、钢、玻璃
	Detail Node 细部节点	

Location: Denver, Colorado, USA
Designer: ZFG Architects LLP
Photographer: Eckert

项目地点：美国科罗拉多州丹佛市
建筑设计：ZFG 建筑师事务所
图片来源：埃克特

Children's Hospital Colorado
科罗拉多州儿童医院

Features 项目亮点

Taking full advantage of natural light, the designer creates a childlike rehabilitation center for pediatric patients through large glass facade with a series of colorful decorates.

设计充分的利用自然光线，通过大面积的玻璃立面与一系列色彩艳丽的空间装饰，为儿童患者创造一个充满童趣的康复中心。

Overview 项目概况

Located on the 160 acre Anschutz Medical Campus at the University of Colorado in Denver, Children's Hospital Colorado has been designed to build an environment that is bright, nurturing, colorful, calm and comfortable; and one that also offers state-of-the-art, world-class care.

儿童医院座落于丹佛市科罗拉多大学占地 160 英亩（约 647 497m²）的安舒茨医学院校园内。设计宗旨是提供一个明亮、呵护、色彩鲜艳、平静和舒适的环境和最先进的世界级医疗护理服务。

The Children's Hospital
Denver, Colorado

LEVEL 1

1 24/7 Fresh Market Place Cafe
2 Boettcher Atrium
3 Chapel and Garden of Hope
4 Creative Play
5 Emergency Department
6 Family Health Library
7 Financial Counseling
8 Gift Shop
9 Main Hospital Entrance
10 Outpatient Laboratory
11 Outpatient Pavilion
12 Outpatient Pharmacy
13 Patient Access
14 Procedure Center
15 Radiology
16 Staff Lounge
17 Volunteers and Patient Reps

Children's ground level plan

Materials and Skin　材料与表皮

The color palette is also of note, consisting of various shades of green, violet, yellow, blue and turquoise, all of which were selected for their healing qualities and applied based on the activities performed in a space. Natural light was also maximized throughout the facility to help with healing.

彩色墙壁也值得一提，各种深浅不同的绿、紫、黄、蓝、绿宝石的颜色都是根据各活动空间和患者康复的需要而选定的。大楼最大程度的自然采光设计有利于患者康复。

Children's elevation south

Children's section east

Function and layout 功能与布局

Integration of Evidence-Based Design principles in the NICU led to additional acoustical protection, lower lighting levels and privacy for visiting families, reflecting the latest medical research that demonstrates low light levels, a quiet atmosphere and privacy improve infants' outcomes. Likewise patient rooms have been designed to host just one child, reducing the rate of infection and external distractions for patients and caregivers. In the spirit of Family-Centered Care, each room has sleeping accommodations for family members, adequate storage, a desk and data ports. Separate family suites have also been included in the facility.

For teenage patients, a 3,000 SF teen-only "hot spot" features a movie theater, pool table, basketball arcade and a music/reading room. Additional indoor play spaces for children are located within the hospital while large, open green spaces are maintained for the same reason outside the hospital walls.

最新的医学研究表明柔和的光线，安静的氛围和私密性有助于提高婴儿的出生率。因此，新生儿重症监护病房增加了隔音、低强度光源、私密空间的设计。与此同时，单人单间的设计以减小患者感染率和被打扰。基于以家庭为中心的护理精神，每个房间都为家庭陪护人员配有休息床位，宽敞的存储空间，办公桌和通迅接口。设计也包括供家庭陪护人员单独使用的套房。

医院配有一个面积为3000平方英尺（约279m²）的青少年患者空间，包括一个电影院、台球桌、篮球场、音乐室和阅读室；也配有儿童活动的室内场所，院外有大型的绿色开放空间。

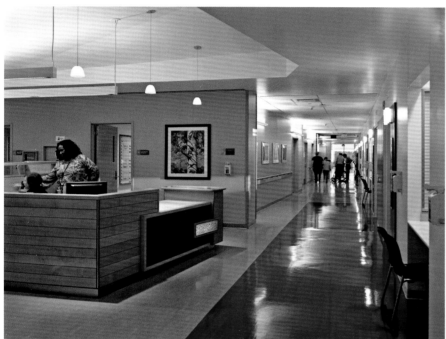

KEYWORD 关键词

Skin and Facade 表皮与立面 | **Glass Facade** 玻璃立面

Materials 材料 | **Glass, Metal** 玻璃、金属材质

Architectural Detail 建筑细部

Location: Haukeland University Hospital, Bergen, Norway
Client: Helse Bergen HF
Architect: C. F. Møller Architects
Engineer: Rambøll
Collaborators: Sweco Norge AS, Multiconsult AS, Conradi Brannteknikk, Kilde Akustikk
Size: 23,000 m²

项目地点： 挪威卑尔根 Haukeland 大学医院
客户： Helse Bergen HF
建筑设计： 丹麦 C. F. Møller 建筑事务所
工程师： Rambøll
合作团队： Sweco Norge AS, Multiconsult AS, Conradi Brannteknikk, Kilde Akustikk
面积：23 000 m²

Laboratories, Haukeland University Hospital
Haukeland 大学医院实验室大楼

Features 项目亮点

The extensive use of glass in the roof, facades and interior walls have minimized the adverse effects caused by the surroudning building block as well as maximized the views and openness.

屋顶、立面和内墙的设计采用了大量的玻璃，既克服了周边建筑遮挡的不利因素，又营造了开放性的视觉空间。

Overview 项目概况

The new Laboratory Building at the Haukeland University Hospital in Bergen is directly connected to the existing hospital, and houses laboratory facilities for both university and hospital use. The new building brings together research laboratories and diagnostic laboratories for the daily work of the hospital. The individual storeys house various kinds of laboratories used in, for example, highly-automated biochemical analysis, microbiology, blood bank functions and genetic research.

The complex is a total of 11 floors, as it was built on top of an existing parking garage. The new building is in very close proximity to the hospital buildings, so that the building's research and diagnostic laboratories act as a liaison and shortcut between the university laboratory buildings to the east and the hospital's patient block to the west. While the proximity is an advantage functionally, it also meant that the building had to be located on a tight and narrow site, facing existing buildings to three sides and a natural rock face to the fourth.

卑尔根 Haukeland 大学医院的新实验大楼直接连接现有的医院建筑，所拥有的实验室设施供大学和医院取用。新实验大楼将研究实验室和医院日常运作的诊断实验室聚合在一起，各个楼层拥有不同种类的实验室，比如高度自动化生化分析、微生物学、血库功能和遗传研究。

建筑总共拥有 11 个楼层，建立在现有的停车场之上。新的建筑非常接近医院建筑，建筑内的研究和诊断实验室成为东边的大学实验楼和西边的医院门诊大楼之间的连接和捷径。虽然地理位置上的接近是一个功能性的优势，但也意味着项目大楼要位于紧凑而狭窄的场地，周围三面被已建成的建筑包围，第四面则是自然岩石。

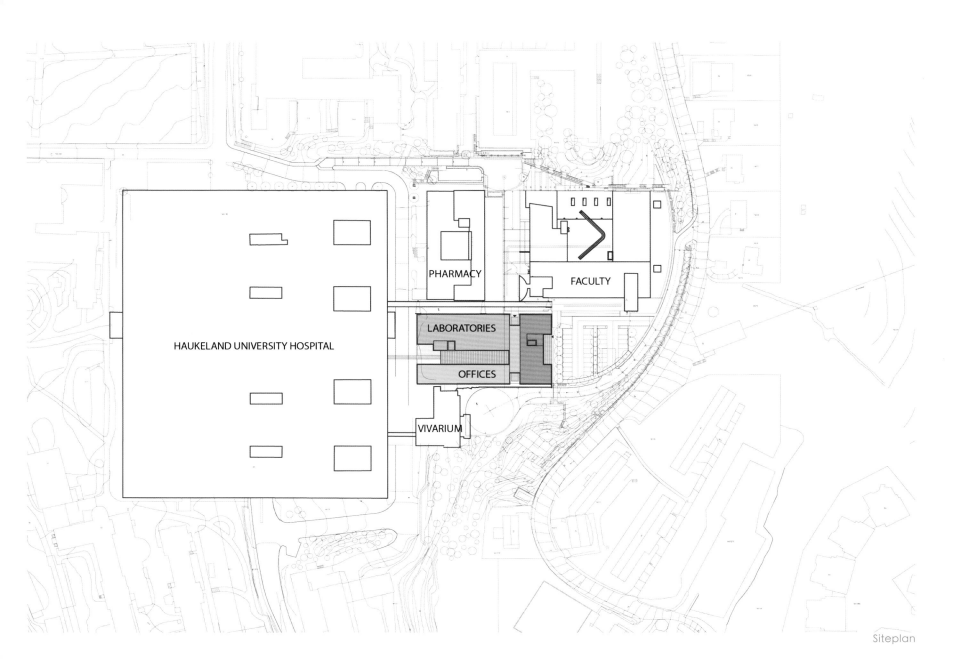

Siteplan

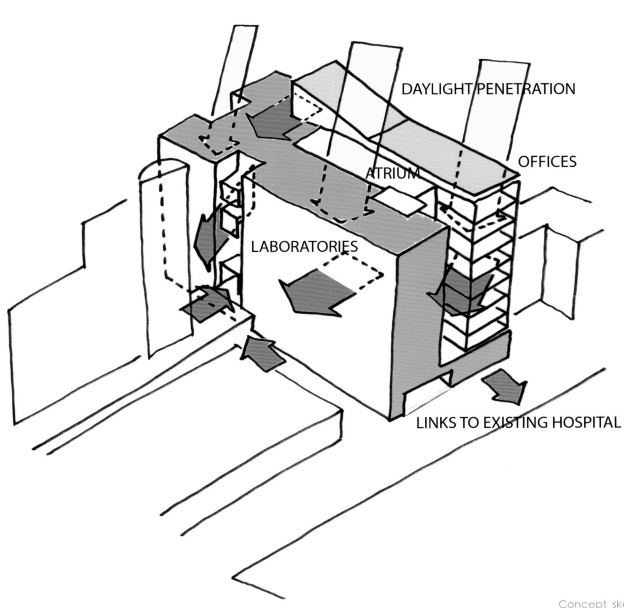

Concept_sketch

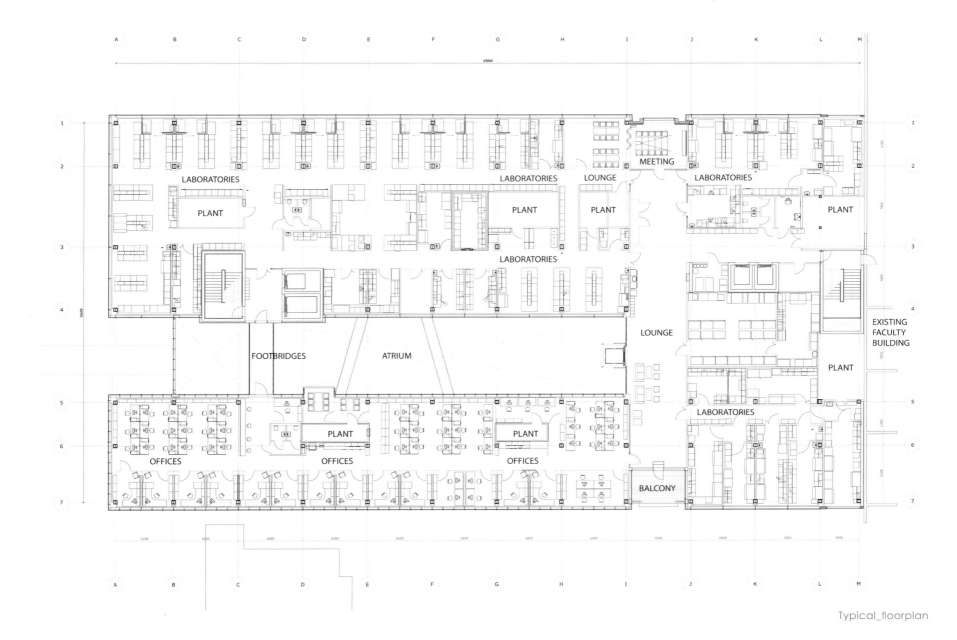

Typical_floorplan

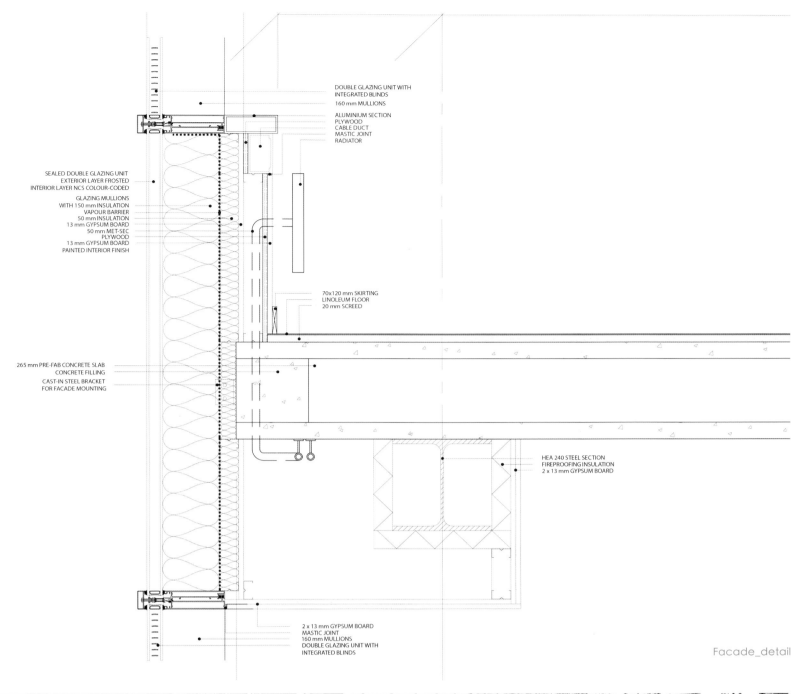

Facade_detail

Facade and Structure 立面与结构

The disadvantages of this constricted, dark location led to extensive use of glass in the roof, facades and interior walls to maximize natural light, views and openness. A nine-storey atrium brings light deep into the building, traversed by footbridges spanning between offices and laboratories. The building meets the high standards of functionality, technology and hygiene in highly flexible laboratories with directly adjacent workplaces. Functional accuracy and clinical purity is also expressed in the precise and colourful glass facades, as a visual counterbalance to the surrounding concrete buildings.

考虑到场地紧凑阴暗的缺点，屋顶、立面和内墙的设计采用了大量的玻璃，以获取尽可能多的自然光和景色，使开放性呈最大化。九层高的中庭将光线引入建筑内部，办公室和实验室之间的天桥横穿过中庭。大楼内部高度灵活的实验室及其他相关临近的工作场所都满足了功能、技术和卫生的高标准要求，功能的准确性和临床纯度还表现在独特多彩的玻璃立面，它使建筑与周围的混凝土建筑保持了视觉的平衡。

KEYWORD 关键词

Skin and Facade 表皮与立面 | **Mixed Facades** 混合立面

Materials 材料 | **Concrete** 混凝土

Symmetry around a Central Longitudinal Axis 纵向轴对称

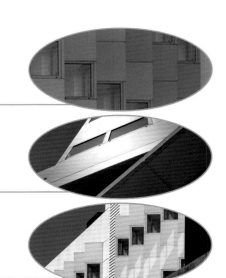

Location: Barcelona, Spain
Developer: FIATC Mútua de Seguros
Architects: JFARQUITECTES
Architect in Charge: Jordi Frontons
Technical Architect: Carlos Arilla
Collaborators: Jaume Serrano Ingeniero de Edificación ; JFG Consultors- Joan Francesc García; Gepro Engineering – Josep Maria Tremps; Genars SL – Francesc Torrebadell / Joaquín Rigau; A3 Arquitectura Técnica-Mª Àngels Sànchez.
Surface: 30,596.44 m²
Photographs: Xavi Gálvez

项目地点： 西班牙巴塞罗那
开发商： FIATC Mútua de Seguros
建筑设计： 西班牙 JFARQUITECTES
技术架构师： Carlos Arilla
合作团队： Jaume Serrano Ingeniero de Edificación ; JFG Consultors- Joan Francesc García; Gepro Engineering – Josep Maria Tremps; Genars SL – Francesc Torrebadell / Joaquín Rigau; A3 Arquitectura Técnica-Mª Àngels Sànchez
表面积： 30 596.44 m²
摄影： Xavi Gálvez

Diagonal Clinic
西班牙 Diagonal 诊所

Features 项目亮点

The main design idea is to create a dialogue between the new building and the surroundings, then conditioned the project by the relation of the different important elements and solved the concrete program.

项目设计的主要理念是通过不同重要因素关系进行项目规划，提出解决方案，建立新建筑与周边的对话。

Overview 项目概况

The new hospital of Mutua FIATC in Barcelona is located in a strategic place of the city, just in Ronda de Dalt west access, one of the most important motorways of Barcelona. The solar characteristics, emphasized by the proximity to the Ronda de Dalt, conditioned the building, which presents a continuous facade opened to Barcelona and remark the outline of one of the main arteries of the city.

The main idea is to create a dialogue between the new building and the surroundings, then conditioned the project by the relation of the different important elements and solved the concrete program.

巴塞罗那的 Mutua FIATC 新医院在城市规划中占据着战略性的地理位置，正好在巴塞罗那最重要的高速公路 Ronda de Dalt 大道西边。建筑的延伸立面面向巴塞罗那，突出了太阳能的设计特点，强调了城市主动脉的轮廓。

项目设计的主要理念是通过不同重要因素关系进行项目规划，提出解决方案，建立新建筑与周边的对话。

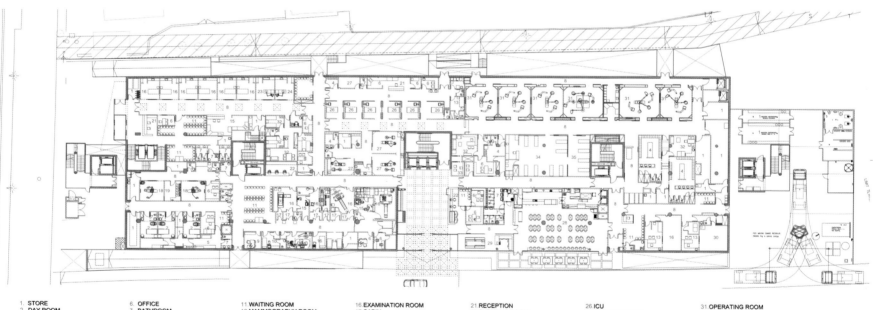

1. STORE	6. OFFICE	11. WAITING ROOM	16. EXAMINATION ROOM	21. RECEPTION	26. ICU	31. OPERATING ROOM
2. DAY ROOM	7. BATHROOM	12. MAMMOGRAPHY ROOM	17. CABIN	22. CLEANING ROOM	27. DELIVERY ROOM	32. STAFF ROOM
3. PREPARATION ROOM	8. CORRIDOR	13. DOCTOR'S OFFICE	18. ENDOSCOPY ROOM	23. PLASTER ROOM	28. BAR-RESTAURANT	33. VISITING ROOM
4. CHANGING ROOM	9. MANAGER OFFICE	14. WAITING BEDS	19. STERILIZATION ROOM	24. HEAL ROOM	29. KIOSK	34. RECOVERY ROOM
5. INFIRMARY	10. SCAN ROOM	15. CONTROL	20. REVIVAL ROOM	25. WORKING ROOM	30. EFFORT TEST ROOM	35. PREANESTHESIA ROOM

◉ FLOOR PLAN

Structure and Materials　结构与材料

The new Diagonal Clinic has 80 rooms, 7 suites, 16 medical offices, 6 operating rooms, 6 emergency boxes and all the services that a hospital need.

The architects propose a monolithic building clearly horizontal, materialized in white precast concrete lied on a concrete podium ink, which allows that the activity might be away of the heavy traffic, becoming apparent with a great noise, and exalt the common areas of the clinic. Like a continuation of the same land, the particular "stylobate", only drilled on the coffee area and on the access, contain the most private uses of the clinic, operating rooms, emergency area, oncology, neonatal... as well as the technical plant of the building.

The upper floors rise up providing consulting areas and rooms with fabulous views of the city. Symmetrically distributed around a central longitudinal axis, rooms are located on both sides, the south side overlooking the city of Barcelona framed by the sea and the Northeast side with the forest of Collserola as a backdrop.

新 Diagonal 诊所设有 80 间病房、7 间套房、16 间医疗办公室、6 间手术室、6 个应急箱，以及医院需要的所有服务设施。

设计师提出独栋建筑的设计，下方是墨色混凝土墩，上方为水平物化的白色预制混凝土，抬高诊所的公共区域，使内部的活动远离外界交通的噪音干扰。犹如在同一片土地上的延续，这个特殊的"柱基"在咖啡区和入口处打孔，划分出诊所中的私人区域、手术室、急诊区、肿瘤科、新生儿区以及技术设备区。

上面的楼层崛起作为咨询区，能够看到城市的美景。各个房间围绕着中央的纵向轴对称分布在两侧，在南边可以俯视巴塞罗那的城市景观，东北面则可以观赏到 Collserola 森林。

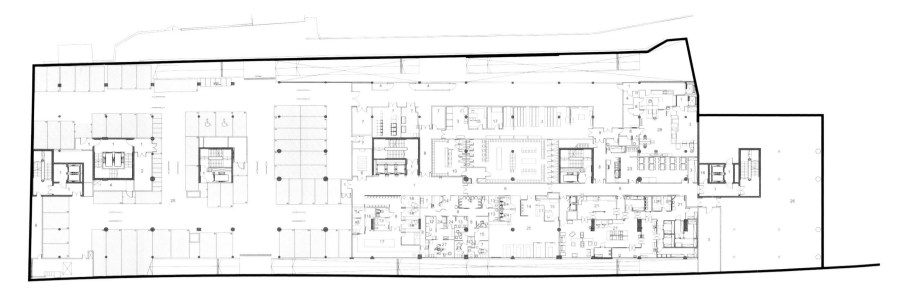

1. LOBBY
2. MORGUE
3. STORE
4. GARVAGE STORE
5. DIRTY CLOTHES
6. ORATORY
7. ROOM
8. CORRIDOR
9. PHARMACY
10. MALE LOCKER ROOM
11. FEMALE LOCKER ROOM
12. WAITING ROOM
13. BATHROOM
14. RECEPTION
15. OFFICE
16. CLEANING ROOM
17. LABORATORY
18. EXTRACTION ROOM
19. GYNECOLOGIST
20. REHABILITATION ROOM
21. KITCHEN
22. LAUNDRY
23. STAF DINNING ROOM
24. CHANGING ROOM
25. PARKING
26. FUTURE EXTENSION
27. OPERATING ROOM
28. PAKAGING AND PREPARATION

BASEMENT PLAN -1

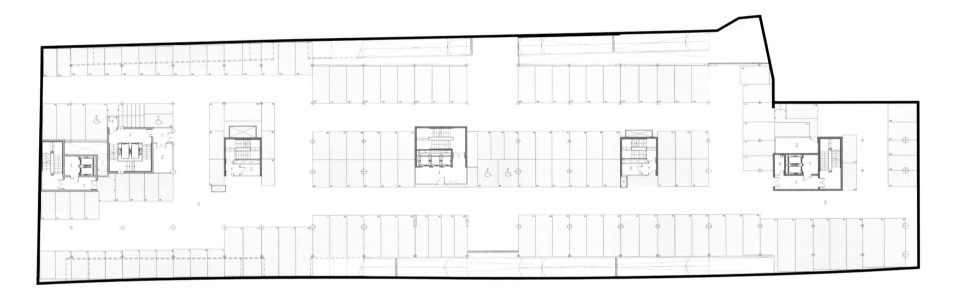

1. LOBBY
2. STORE
3. PARKING
4. CLEANING ROOM

BASEMENT PLAN -2

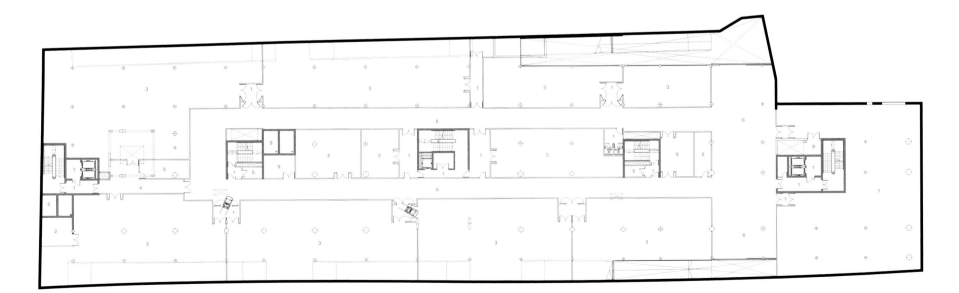

1. LOBBY
2. TECHNICAL ROOM
3. ARCHIVE
4. HALL
5. STORE
6. CORRIDOR
7. FURUTE ADDITION
8. BATHROOM
9. TANK-STORE

BASEMENT PLAN -3

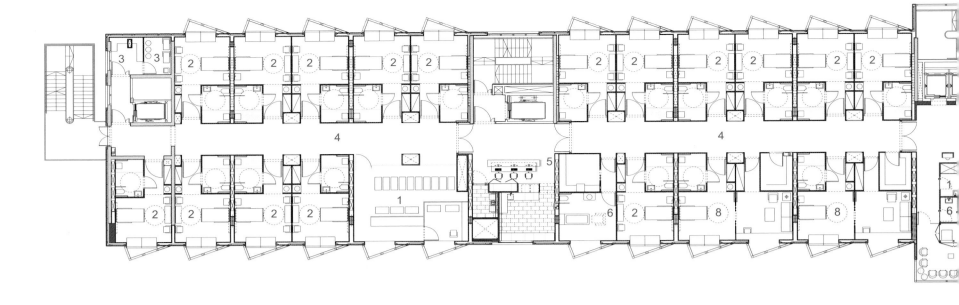

1. NURSERY
2. ROOM
3. STORE
4. CORRIDOR
5. RECEPTION
6. BATHROOM
7. RACK
8. SUIT ROOM
9. VIP WAITING ROOM
10. LOBBY

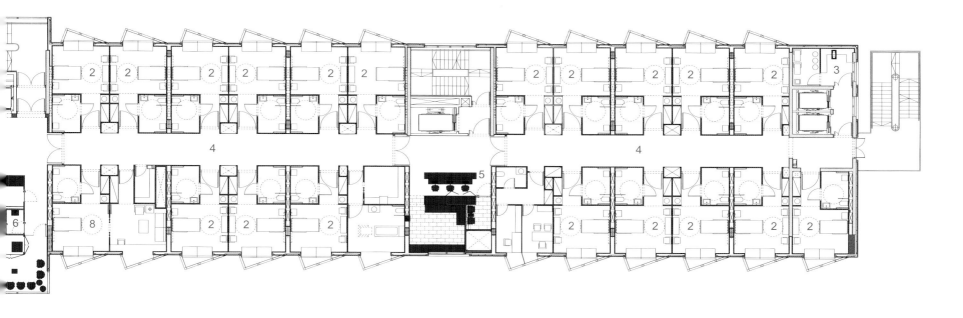

Room floor plan.pdf

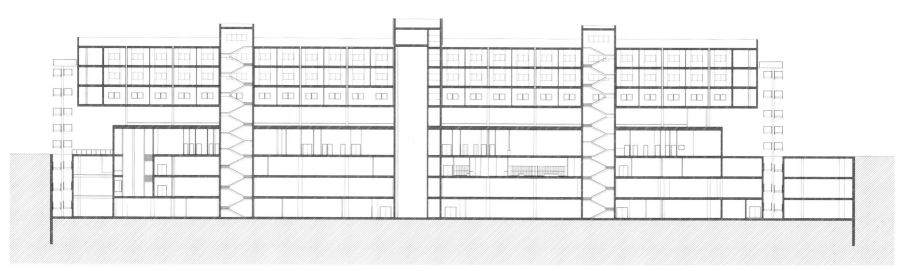

section 2

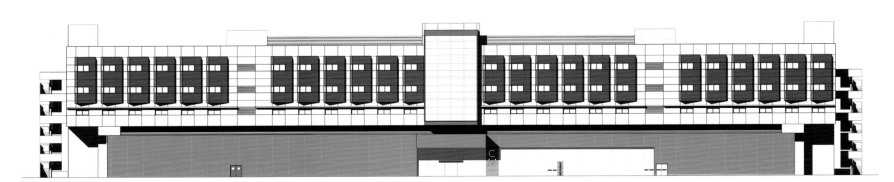

Elevation

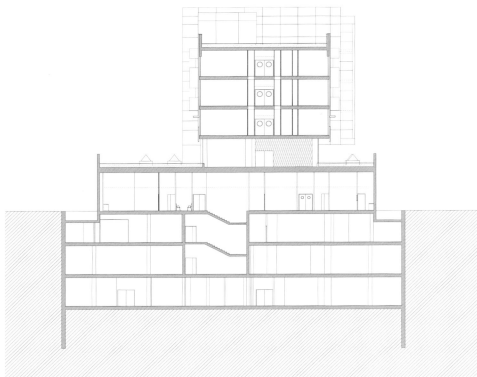

section 1

KEYWORD 关键词

Skin and Facade 表皮与立面 | **Mixed Façade** 混合立面

Materials 材料 | **Glass, Ceramic Tile, Metal** 玻璃、瓷砖、金属

Cross-shaped 十字造型

Location: Kochi, Japan
Developer: Kochi Health Sceiences PFI
Architectural design : AXS Satow Inc. Nishikawa Sekkei
Landscape design: AXS Satow Inc. Nishikawa Sekkei
Land Area: 72,092 m²
Total Floor Area: 67,396 m²
Building-to-land ratio: 23.3%
Floor-area ratio: 93%

项目地点：日本高知县
开发商：高知县卫生科学系统
建筑设计：佐藤综合计画及西川设计
景观设计：佐藤综合计画及西川设计
占地面积：72 092 m²
建筑面积：67 396 m²
绿化率：23.3%
容积率：93%

Kochi Health Sciences Center
高知健康科学中心

Features 项目亮点

Use glass, light metal and ceramic title as skin's main materials. And the cross-shaped upper floor not only enriched the building form but also create favorable landscape view.

建筑采用玻璃、轻质金属以及瓷砖等作为表皮的主要构造，十字形的上层楼面造型既丰富了建筑形式，又创造了良好的景观视线。

Overview 项目概况

As Kochi Prefecture's foremost medical center, Kochi Health Sciences Center is entrusted with assisting remote communities with their medical needs and providing emergency services during natural disasters. Through ongoing maintenance of its advanced emergency medical center function, advanced medical information system, and day surgery services, the hospital provides patients with a safe, comfortable medical environment. Aiming to improve the efficiency of its medical services, Kochi Health Sciences Center has become the first hospital project in Japan developed in the framework of the Private Finance Initiative (PFI).

高知健康科学中心作为高知县最重要的医疗中心，承担着当地社区的医疗需求以及在自然灾害中提供急救服务的重任。医院通过其先进的急救中心功能，先进的医疗信息系统以及全天候手术服务为病人提供了一处安全舒适的住院环境。高知健康科学中心旨在提高医疗服务的有效性，是私人融资机构在日本投资的第一家医院。

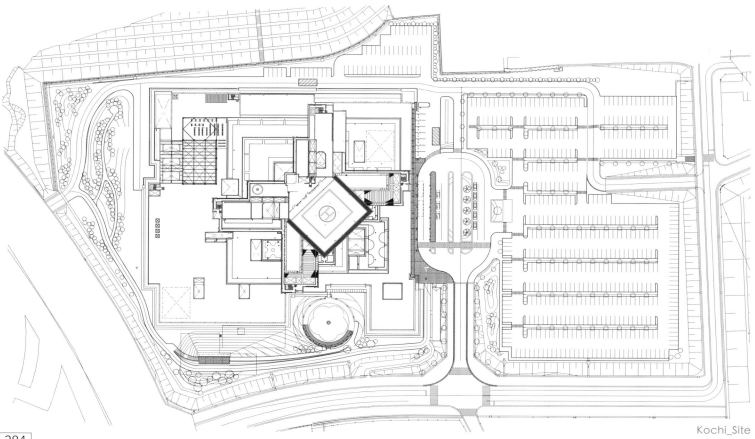

Kochi_Site

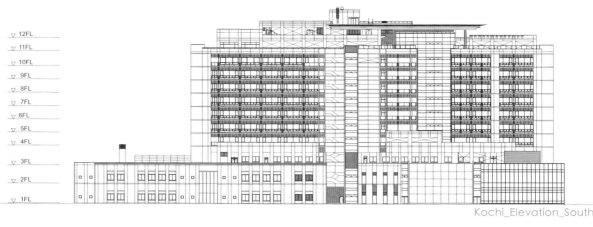

Kochi_Elevation_South

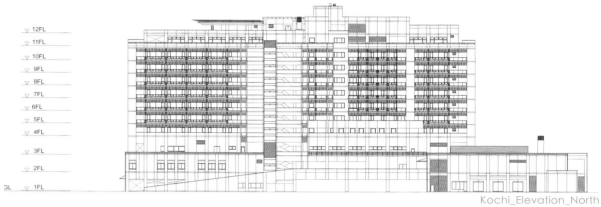

Kochi_Elevation_North

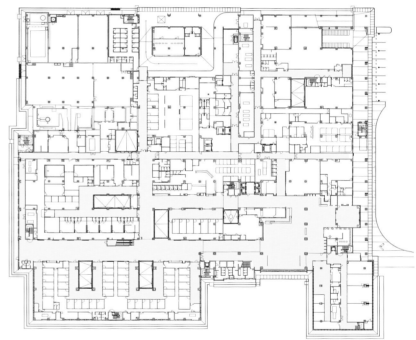

Kochi_Plan_01F

Structure and Shape 结构与造型

In designing the hospital, the designers chose to exclude a basement level and thereby cut costs, shorten the construction period, and minimize the volume of surplus earth. To distance the hospital rooms from an artificial mountainside which looms oppressively, north of the site, we laid out the upper floors on a cruciform plan. In support of the Center's advanced medical services, the design maintains sufficient interior space for diagnosis, nursing, and bedside rehabilitation. Consideration has also gone to patient amenities, such as hospital rooms of open design allowing patients to have exterior views while lying in bed.

设计该医院的过程中，设计师没有设计地下室以控制成本和缩短建造时间，同时使空余土地面积最小化。为使医院病房离人造假山的距离远一点，设计师将建筑北边的上层楼面设置成十字形。为满足医院先进的医疗服务，设计师更多地考究了诊疗室的空间设计，看护病房空间以及住院部空间。同时更多地考虑到病人的感受，如开放式的病房设计使病人摊在床上便能欣赏医院外部景致。

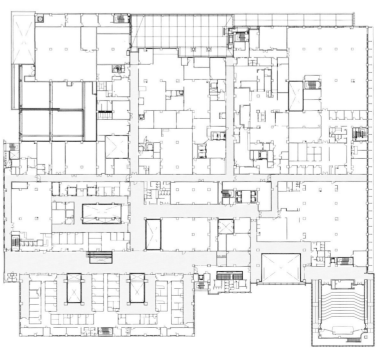

Kochi_Plan_02F

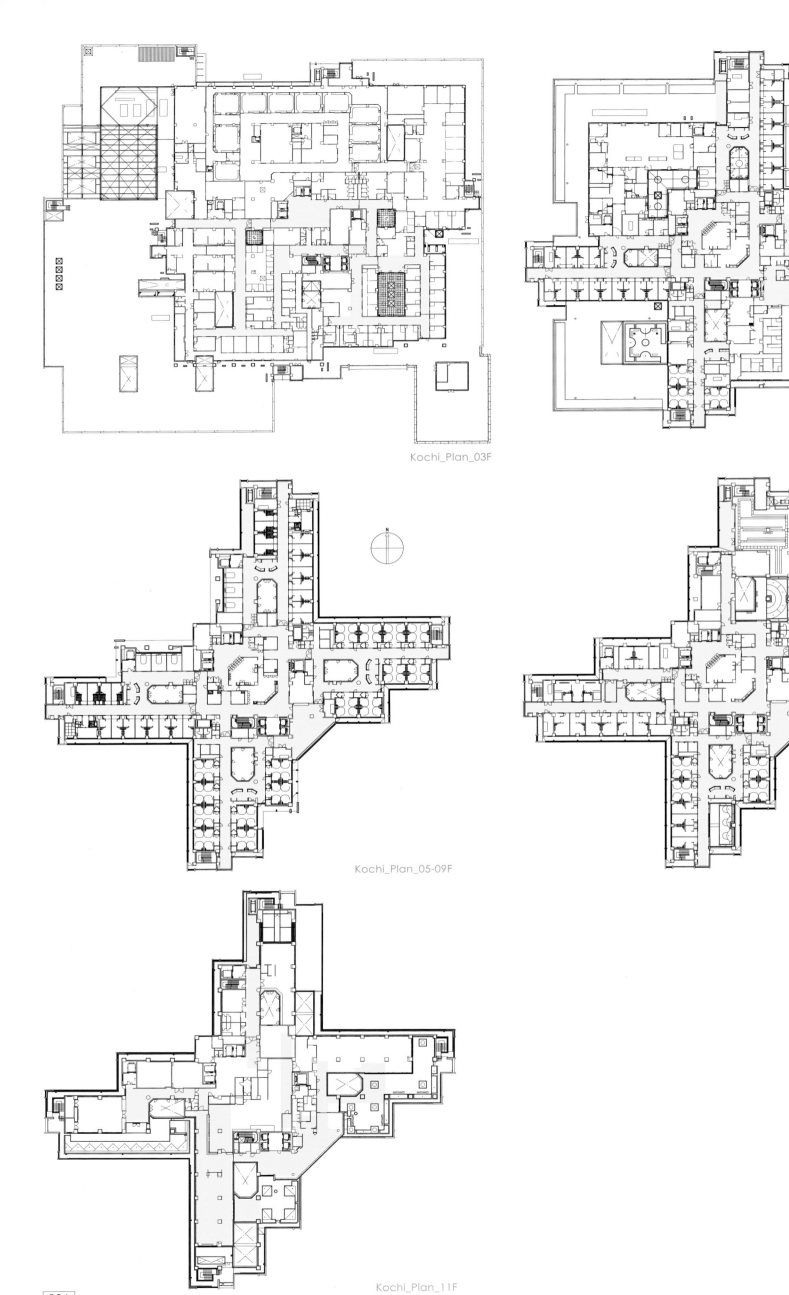

Art Feeling
Architectural
Implication
Abstract Form
Innovative Structure

艺术内涵
建筑意蕴
抽象形式
创新结构

Art Building
艺术建筑

With the development of modern art, buildings for this subject are also developed a lot. Art buildings generally include display and exhibition buildings for art items and public buildings for cultural and art communications, namely, museum, art gallery, concert hall, theater, etc. Art itself is a kind of ideology to satisfy spiritual needs, thus art building should be symbolic or monumental to deliver the cultural meaning it bears.

From facade, color and eave, to topography, landscape, waterscape and cultural connotation, the design of art buildings pays attention to every detail of the space: comparison and changes, rhythm and scale, connection and transition, penetration and levels, guidance and implication, etc. In terms of architectural form design, it's gradually changed from traditional architectural style to modern abstract style. Roof structure, wind-resistant design, earthquake-resistant design, foundation design and anti-corrosion design are greatly improved. Meanwhile, building materials are carefully selected according to different functions. For example, the concert hall emphasizes sound effect, thus the materials should be selected to enhance this point.

随着当代艺术的发展，与之相适应的艺术类建筑也得到长足发展。艺术类建筑通常指陈列、展览有关艺术方面的事物，提供文化艺术交流场所的公共建筑，如博物馆、美术馆、音乐厅、剧院等。艺术本身是满足人们精神需求的意识形态，因而，各种类型的艺术建筑，也通常离不开其所承载的文化象征意义、纪念意义，并成为整个时代和文化的重要表征。

艺术建筑从立面、色彩、屋顶檐口的设计，到基地内外的地形地貌、景观配置、水系利用和文化底蕴等方面均追求空间的对比与变化、韵律与节奏、比例与尺度、衔接与过渡、渗透与层次、引导与暗示。造型上，艺术建筑经历了从传统建筑样式到抽象建筑样式的转变。屋盖结构的建模、抗风设计、抗震设计、基础选型、防腐蚀措施等结构设计也有了很大的创新。同时根据不同的功能用途，如音乐厅就非常注重其音乐声响效果的把握，建筑材料选择上也会有很大的差异。

KEYWORD 关键词		
Skin and Facade 表皮与立面	Wooden Facade 木质立面	
Materials 材料	Concrete, Steel, Glass, Wood 混凝土、钢、玻璃、木材	
Landscape Context 景观文脉		

Location: Alise-Sainte-Reine, France
Architect: Bernard Tschumi Architects
Site Area: 370,000 m²
Floor Area: 8,000 m²
Maximum Building Height: 18 m, wood screen (parapet) is 13 meters high.
Photography: Christian Richters, Iwan Baan

项目地点：法国阿利兹-圣-莱纳
建筑设计：法国 Bernard Tschumi 建筑师事务所
占地面积：370 000 m²
建筑面积：8 000 m²
建筑高度：18m（木墙13m）
摄影：Christian Richters, Iwan Baan

Alésia Archaeological Museum
阿莱西亚考古博物馆

Features 项目亮点

The new museum complex recreates battlements and earthworks and provides interpretation for the area, which consists of several sites spread over a valley that contains a small medieval town. And both of the two objects appear as non-obtrusive as possible in their respective contexts.

新建筑通过表皮材料来适应周边环境，造型上也与当地的建筑相仿，为人们再现了中世纪时期的城堡和小镇的一些意象。

Overview 项目概况

The project marks an archeological site in central France and commemorates the history of the battle between Julius Cesar and the Gauls in 52 B.C. Although all traces of the battle have been obliterated, the new museum complex recreates battlements and earthworks and provides interpretation for the area, which consists of several sites spread over a valley that contains a small medieval town.

这座位于法国阿利兹-圣-莱纳的博物馆是为了纪念公元前52世纪末朱利叶斯凯撒与高卢人之间的战争而建。战争遗迹已随着岁月流逝，而这座新建筑的形象则为人们再现了中世纪时期的城堡和小镇的一些意象。

PLAN MASSE

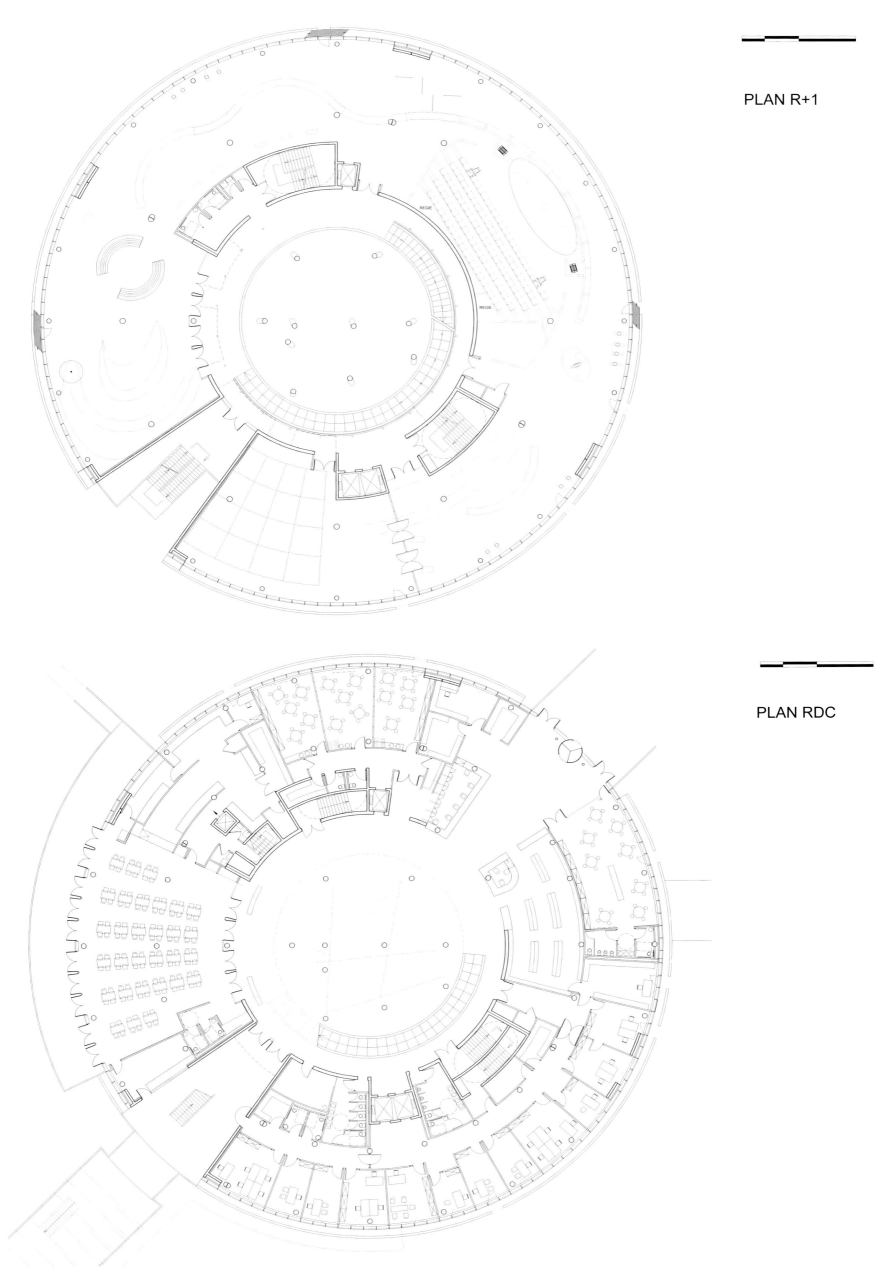

PLAN R+1

PLAN RDC

CROQUIS

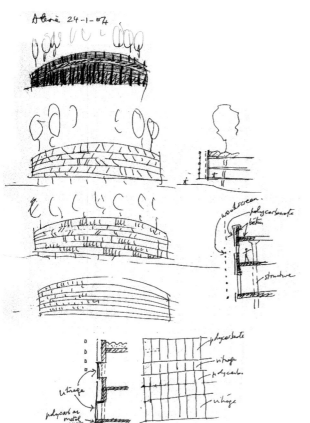
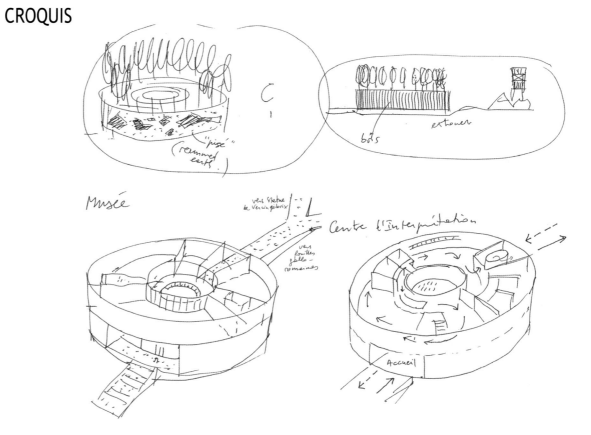

Structure and Shape 结构与造型

The scheme consists of two separate but related structures. One building is a museum, located at the position of the Gauls during the siege at the top of the hill above the town. A second building is a visitors' center located at the Roman position in the fields below the town. The public client wanted both buildings to appear as non-obtrusive as possible in their respective contexts. The museum is built of stones, similar in look to the town buildings, but with contemporary technology, and is buried partially into the hill so that from above it appears as an extension of the landscape. Visitors may go to the roof to view the surrounding landscape from the position that the Gauls did two thousand years ago.

建筑由两个独立而相互关联的部分组成。博物馆位于当年高卢人被包围的小山顶上，游客中心位于当年罗马阵营所在的位置。客户希望两座建筑都不要显得过于突兀，要能够融合在各自的环境背景中。博物馆由石头建造，形态与小镇里的建筑相仿，但是所使用的建造技术是现代的。博物馆的一部分被埋入地下，因而从远处看去，成为此处景观的一个组成部分。游客可以登上屋顶，从两千年前高卢人的视角去观赏周围的景色。

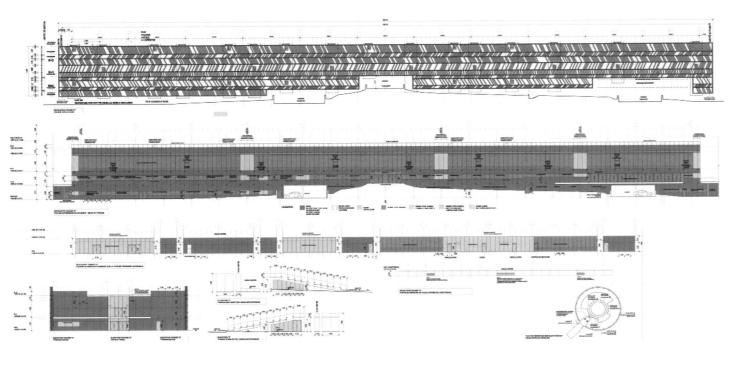

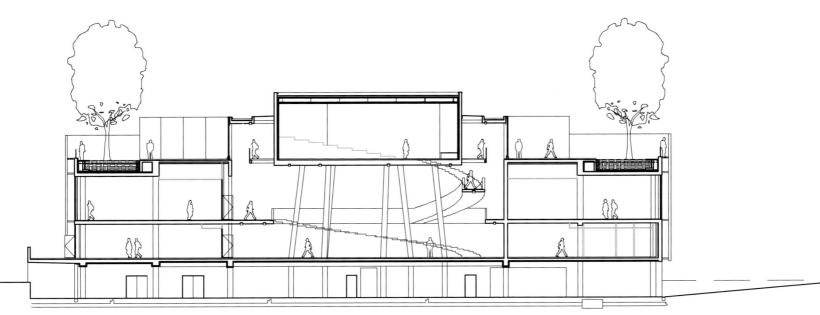

Sectional Drawing 剖面图

Skin and Materials 表皮与材料

The interpretative center is built of wood, much as the Roman fortifications would have been at the time of the siege. The roof of the building is a garden planted with trees and grass, camouflaging the presence of the building when seen from the town above. Visitors may look onto reconstructions of the Roman battlements from the roof garden, or stroll down a path to experience the reconstitutions first-hand. A keen awareness of the surrounding landscape as it pertains to the historic battle is integral to the visitors' experience. The context of the site is primarily the natural, verdant landscape of Burgundy and the medieval buildings of the town of Alise-Sainte-Reine. A 360° panoramic view is required by each of the buildings. The envelopes adapt to their surroundings through materials, while the form of the buildings is deemphasized.

游客中心是由木头建造的，屋顶种植绿化，从高处的小镇看去，它就像一座花园，游客可以站在屋顶或漫步于环坡道，体会置身罗马城堡中的感受。通过周围景观再现历史是游客体验的一个重要组成部分。勃艮第地区的自然景观以及中世纪小镇风貌是这里的景观文脉，简单而抽象的形体激发游客对于历史的尊重和敬畏。建筑的形式被弱化，并通过表皮材料来适应环境。

DETAILS

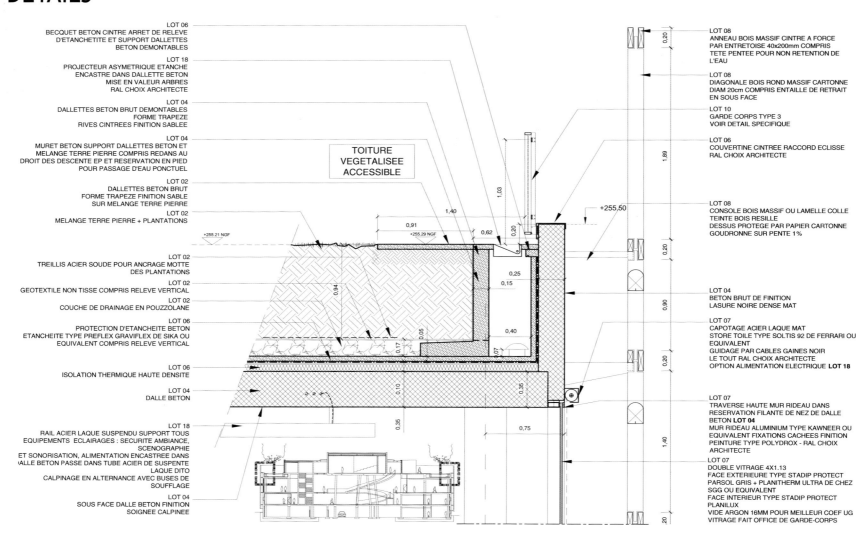

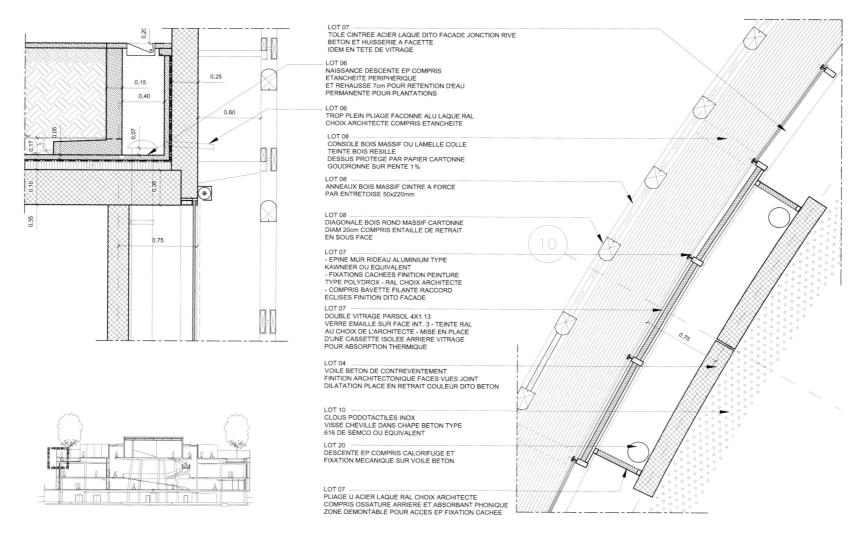

| KEYWORD 关键词 | Skin and Facade 表皮与立面 | Stone Facade 石材立面 |

| | Materials 材料 | Limestone, Miramichi Sandstone, Stained Glass 石灰岩，米拉米奇砂岩，彩色玻璃 |

| | Architectural Style 建筑风格 | |

Location: Montreal, Quebec, Canada
Client: Montreal Museum Fine Arts
Architectural Design: Provencher Roy + Associés Architectes
Design Team: Claude Provencher, Matthieu Geoffrion, Eugenio Carelli, Jean-Luc Rémy, Denis Gamache
Area: 5,483 m²
Photography: Tom Arban, Alexi Hobbs, Marc Cramer, Jean-Guy Lambert

项目地点：加拿大魁北克蒙特利尔
客户：蒙特利尔艺术博物馆
建筑设计：Provencher Roy + Associés Architectes
设计团队：Claude Provencher, Matthieu Geoffrion, Eugenio Carelli, Jean-Luc Rémy, Denis Gamache
面积：5 483 m²
摄影：Tom Arban, Alexi Hobbs, Marc Cramer, Jean-Guy Lambert

A Canadian Museum in a Church
Claire and Marc Bourgie Pavilion of Quebec and Canadian Art
蒙特利尔艺术博物馆：古老教堂中的艺术博物馆

Features 项目亮点

Transition and continuation methods are meticulously used in designing this museum, the new buildings integrated into old buildings and its surroundings spontaneously.

该博物馆的设计以巧妙的过渡、延续手法，将新建筑自然地融入旧建筑中，并使得新馆与周围环境完美融合。

Overview 项目概况

Beside the restored former Erskine and American Church, transformed into a 444-seat concert hall, the rear annex has been completely reconstructed in contemporary style to form the new art pavilion. "The project was complex because it entailed designing a building capable of featuring the Quebec and Canadian art collections while establishing a dialogue with the church, with the museum's other pavilions, and with the city," explained Claude Provencher, founding partner of Provencher Roy + Associés Architectes.

With its restraint and permeability, the new Claire and Marc Bourgie Pavilion of Quebec and Canadian Art establish a natural dialogue with the city. From every level, the glazed openings offer a view of the city and, at the building's foot, the museum's sculpture garden, a linear exhibition of works of public art bordering the museum. In addition, the glassed-in atrium at the top of the pavilion offers a strong visual link with Mount Royal, an emblematic element of Montreal's identity. Another dialogue is established with the church. The new pavilion shares more than its entrance and reception areas with the former religious building. It also evokes its spirit both by its elevation, extended by an opening to the sky, and by the presence of subtly designed alcoves around the galleries.

The central stairway, providing a link between the chiaroscuro of the basement and illumination at the top of the building, allows for this crescendo of light (despite the moderated lighting that is required in most of the galleries for conservation of the artworks) with progressively larger and larger visual openings to the

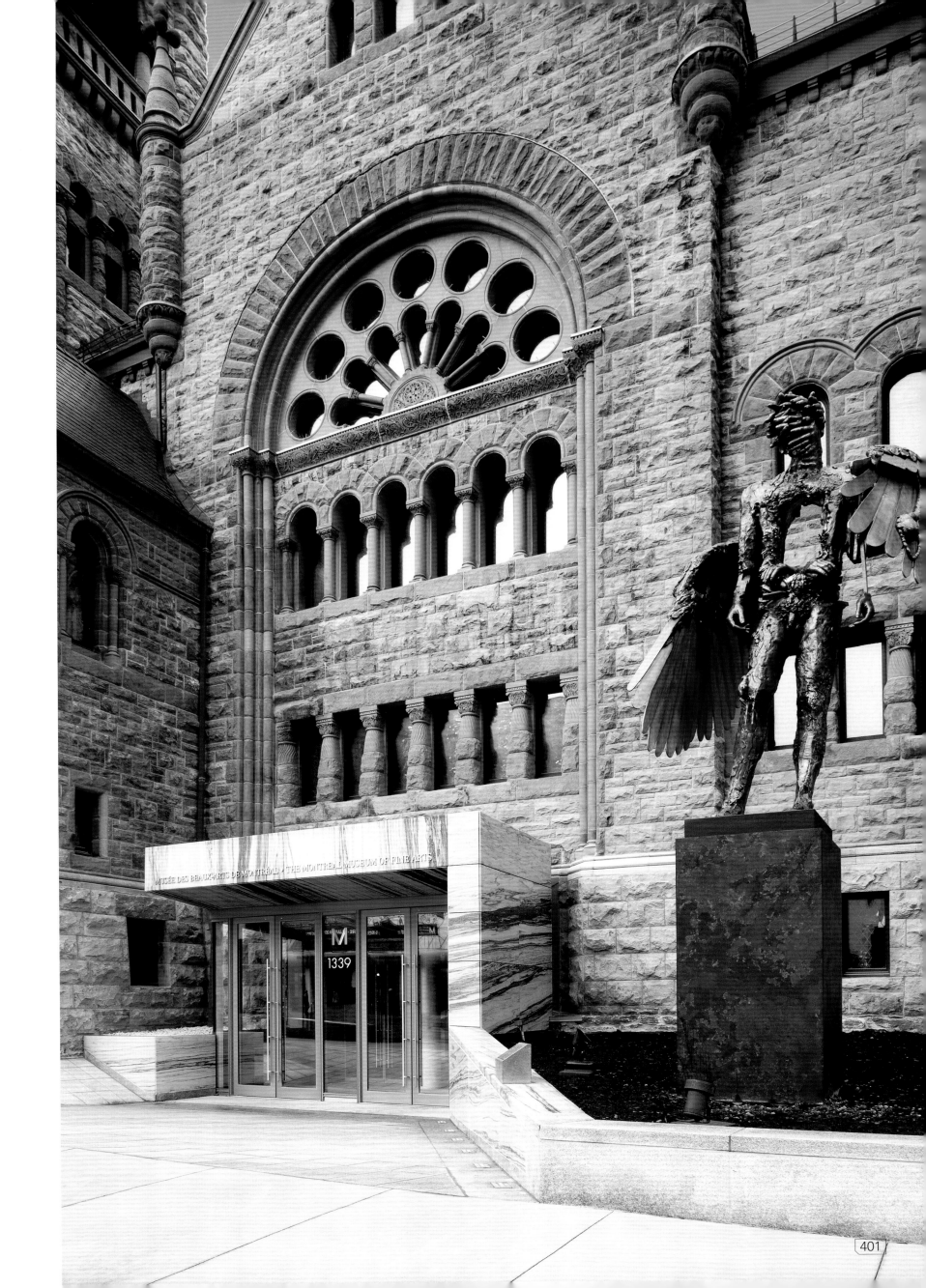

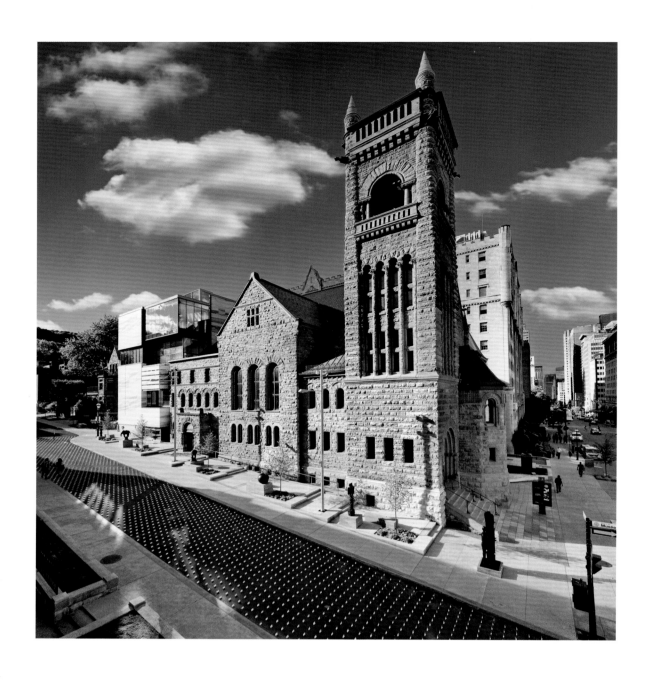

exterior. On the third level, a bay window gives onto an outdoor terrace that features the new work by Dominique Blain. Aside from its emphasis on light, the glassed-in atrium on the top floor has a figurative value. Its shape, evoking an ice structure inspired by an igloo, establishes a symbolic link with the Inuit works that are on display below.

蒙特利尔艺术博物馆决定保留尼斯金美式教堂的完整性，经过修缮将其转化为一个 444 座的音乐厅。设计师认为"该项目的挑战在于设计一个可以与古老教堂自然共存的加拿大魁北克特色艺术收藏品博物馆，并与城市交相辉映"。

项目依据老建筑的外观元素进行巧妙的延续，每一层的玻璃开口提供了良好的景观视野，让新建筑与城市建立对话，并让新老建筑的结合自然过渡。建筑周围的地面作为博物馆雕塑品公园，线性陈列的雕塑作品创造了公众艺术品空间。展馆顶层中庭的大型玻璃幕顶提供了建筑与作为蒙特利尔身份象征的 Mount Royal / 皇家山的视觉联系。另一种对话建立在新艺术馆与旧教堂之间，新老建筑之间不仅仅共享了入口处和接待区的空间，他们的精髓也在相互影响，包括立面的衔接，统一的开放性甚至是新建筑画廊中设计巧妙带有宗教色彩的壁龛。

中央楼梯为建筑较暗的底层和明亮的顶层提供连接，除了渐强的光线（尽管出于保护艺术品的考虑，大多数画廊的照明应该是幽暗的），建筑也提供了逐渐开阔的外部视野。不仅仅是强调轻盈，顶层的玻璃中庭作为一种比喻，它的形状灵感来自于冰屋结构，这也让建筑结构与下层展示的因纽特人作品产生符号化的关联。

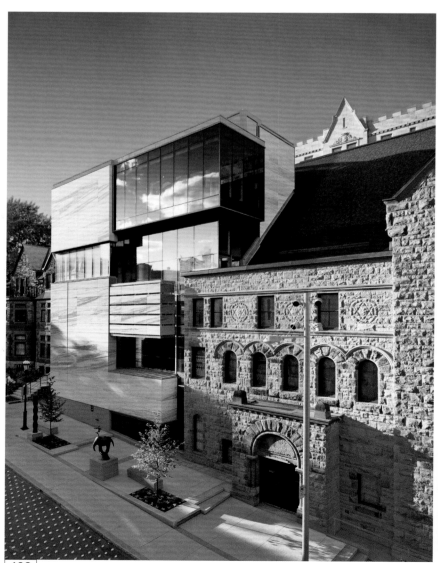

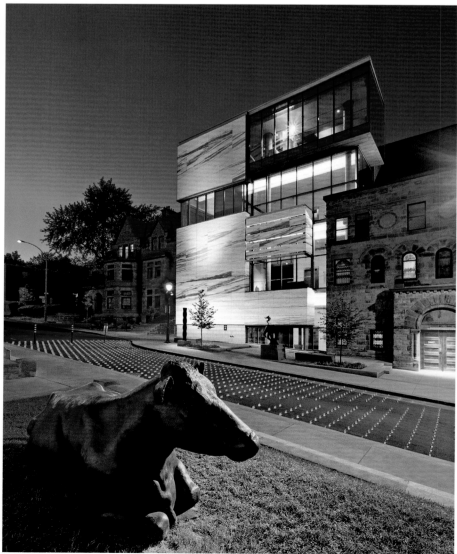

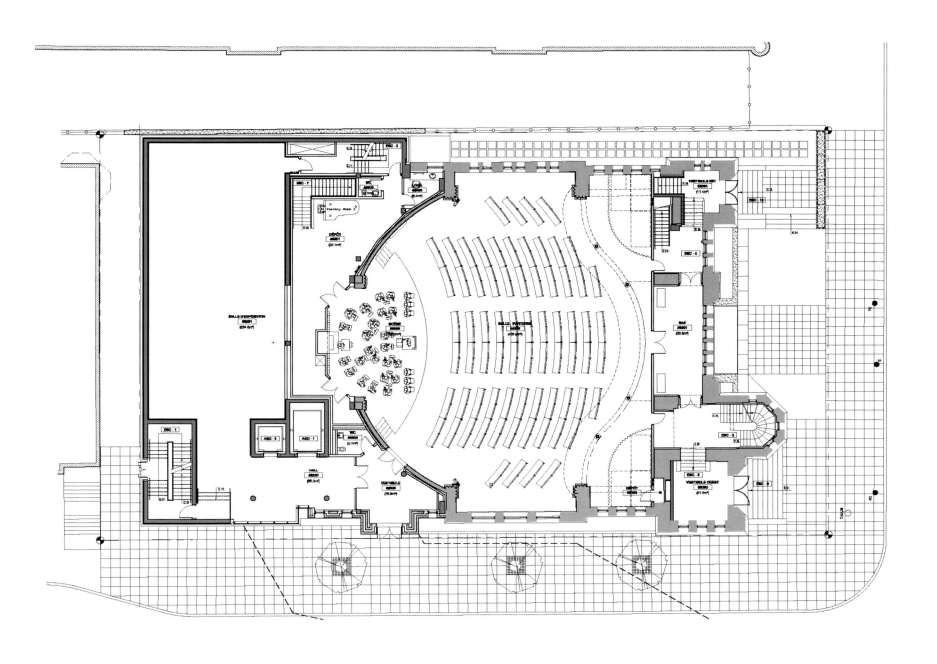

Shape and Materials 造型与材料

Built in the late nineteenth century in the massive neo-Roman style of the Trinity Church in Boston, the Erskine and American Church is an important patrimonial landmark, not only for its architecture but also for its history and its contribution to Montreal's urban landscape. Evidence of the rise to power of the city's Scottish Protestant élite and an era when the "Golden Square Mile" was home to 70% of Canadian wealth, this building, made of limestone with insertions of Miramichi sandstone, has an original textured façade and a Byzantine-style dome testifying to the city's architectural richness. In addition, the 20 Tiffany stained-glass windows gracing the lateral façades form the largest collection of its type in Canada.

　　厄斯金美式教堂作为蒙特利尔市重要的传统地标，除了建筑风格外，更重要的是其历史意义以及为城市景观带来的巨大贡献。像在蒙特利尔具有重要历史意义的"黄金广场区"中许多建于十九世纪末二十世纪初的石制建筑一样，厄斯金美式教堂以石灰岩混合米拉米奇砂岩建造，它以原始的外貌与拜占庭式拱顶见证了城市建筑风格的多样化。此外，20块帝凡尼彩色玻璃窗使其成为加拿大拥有最多帝凡尼玻璃窗的建筑。

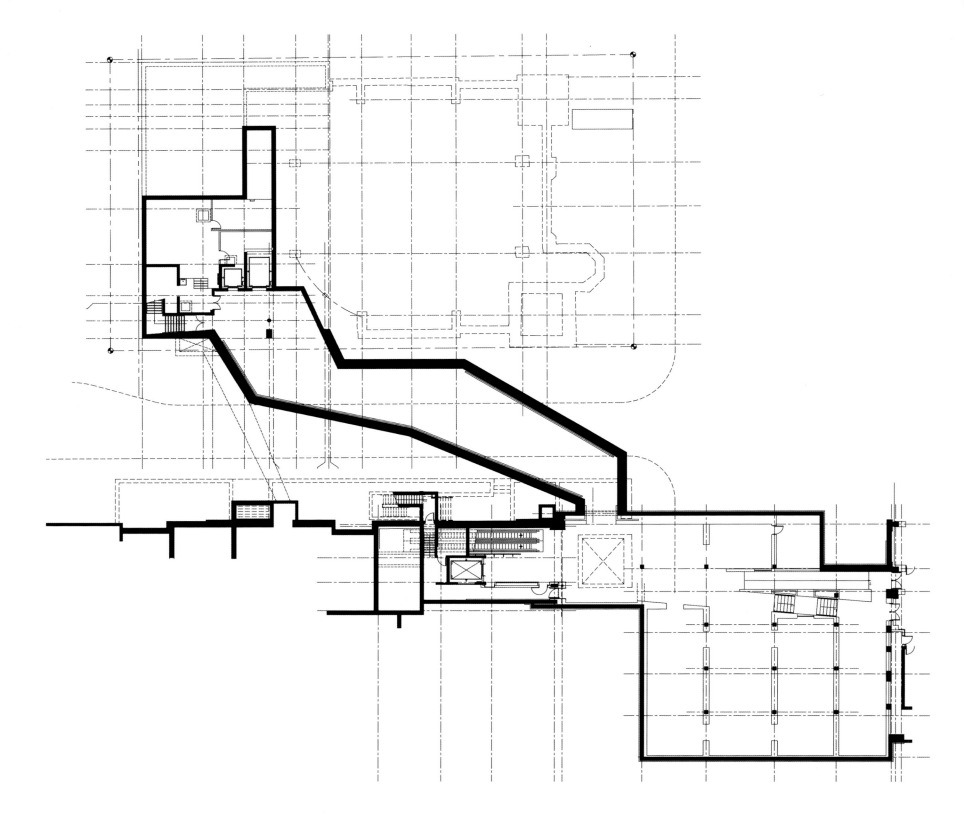

Skin and Facade 表皮与立面

The architects of Provencher Roy + Associés Architectes meticulously restored the church's envelope. Some parts that were too badly damaged were completely resculpted. The 146 stained-glass windows (including the 20 Tiffany windows) were removed, restored, and reinstalled behind glass panels that make the building watertight. Inside, the plaster ceiling and the mouldings were completely restored, as were the woodwork pieces adorning the nave.

The architects clad the new pavilion with a wall of marble from the same Vermont quarry as that of its two predecessors, reproducing the image of the material in its original state, with the design of veins running along the façade. "We wanted to give the feeling that the galleries had been sculpted from a gigantic four-storey-high block of marble," recalls Matthieu Geoffrion, project manager for the Bourgie Pavilion.

建筑师精心策划教堂外立面的修缮，一些损坏严重的部分被完全修复。146个彩色玻璃窗（包括20个帝凡尼玻璃窗）被拆除，修复并重新安装到具有防水性能的玻璃板背后。在建筑内部，中殿的石膏天花经过修复并增加了木件装饰。

建筑师在新建筑外立面重新诠释了白色大理石材质的应用，这也象征着新馆作为集成性综合艺术馆的功能。建筑师和两任前辈一样，采用Vermont采石场的大理石，让同样的纹理作为设计脉络，在新老建筑间流传。

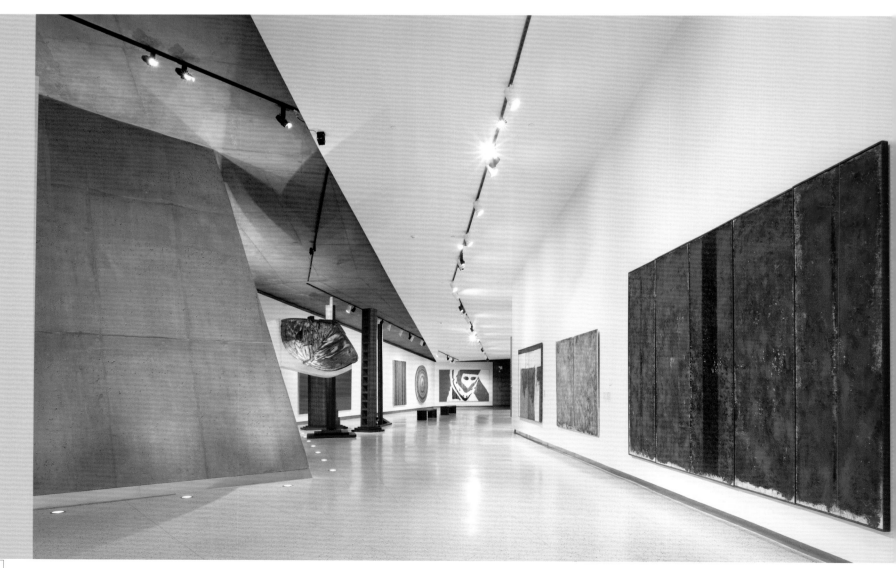

KEYWORD 关键词

- **Skin and Facade** 表皮与立面 | **Glass Facade** 玻璃立面
- **Materials** 材料 | **Glass** 玻璃
- **Curved Canopy** 弧形檐篷

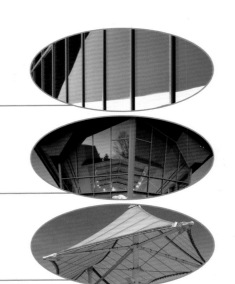

Location: Oklahoma City, Oklahoma, USA
Client: The National Cowboy and Western Heritage Museum
Architect: Fentress Architect
Size: 22,297 square meters

项目地点：美国俄克拉何马州俄克拉何马城
客户：美国西部牛仔历史博物馆
建筑设计：芬特雷斯建筑事务所
总建筑面积：22 297 m²

National Cowboy and Western Heritage Museum
美国西部牛仔历史博物馆

Features 项目亮点

After this major expansion, the new entry to the museum is anchored by a sweeping curved canopy. This layout allows the facility an excellent forum for showing and integrating art with its architectural surroundings.

在改造的过程中，扩大了面积，并在博物馆入口处设计了弧形檐篷。艺术与建筑环境更好的结合使这里变成了一个极佳的交流场所。

Overview 项目概况

The National Cowboy and Western Heritage Museum has put its own brand on the world of cultural institutions. Its subject is the working cowboy, from the Civil War era to today and its mission is to collect, preserve and interpret objects of all sorts that refer to this archetypal inhabitant of the American West.

The museum, founded in 1955 in Oklahoma City by a group of art and history lovers representing more than a dozen western U.S. states, needed more room and better organization to showcase its collections, events and experiences, including an unusual—and sizeable—replica of an Old West town.

作为世界文化机构的一员，美国西部牛仔历史博物馆陈列对象是从美国内战至今的牛仔。该馆收集、保存和诠释牛仔——这个最具美国西部典型群体的所有相关物品。

博物馆位于俄克拉何马城，由一群来自美国西部各个州的艺术和历史爱好者于1955年组织成立，目的是更好的展示牛仔们的物品、事件等。这些物品当中包括一个非常难得的、巨大的西部旧城模型。

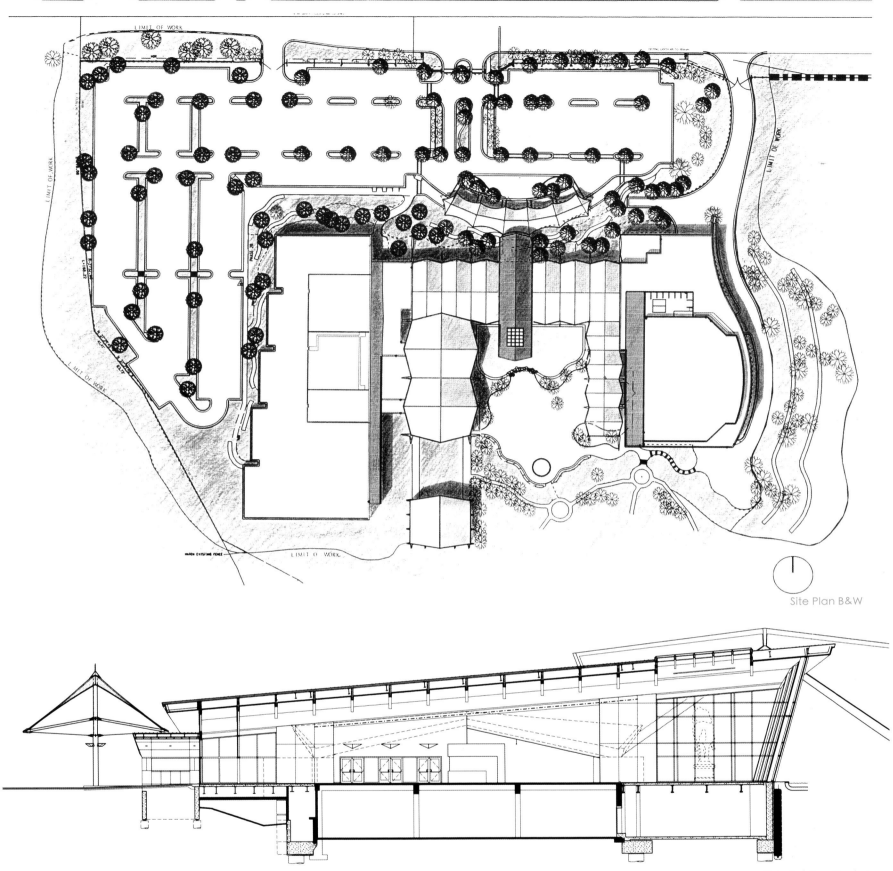

Site Plan B&W

Bldg Section flipped

Structure and Shape 结构与造型

Prior to this major expansion, the Museum consisted of a two-level, 7,153 square meter gallery, exhibit, and administrative facility. The now 22,297 square meter museum allows for the rearrangement of gallery and curatorial spaces, placing all gallery space on the main level, and the curatorial and storage functions on the lower level. This layout allows for the addition of over 6,968 square meter of new state-of-the-art flexible gallery and exhibit space, making this facility an excellent forum for showing and integrating art with its architectural surroundings.

The new entry to the museum is anchored by a sweeping curved canopy, which gathers people in a welcoming gesture. This architectural gesture reinforces a sense of entry, lost in the original plan, which aligns the visitor with one of the strengths of the existing complex, the pool and sculpture garden beyond.

原博物馆包括画廊、展览馆和行政处，共有两层，总面积 7153 m²。扩建后，博物馆总面积达到 22 297 m²。新的画廊、展览馆空间内加了一个存储间，总面积比原来增加了 6 968 m²。艺术与建筑环境更好的结合，使该博物馆成为一个极佳的交流场所。

博物馆入口处的弧形檐篷，像在欢迎来此参观的人们。改造后的入口比原来的入口特征更为鲜明，并且能将博物馆、水池及雕塑花园与游客连在一起。

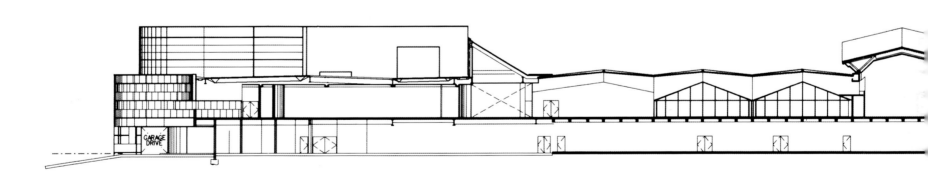
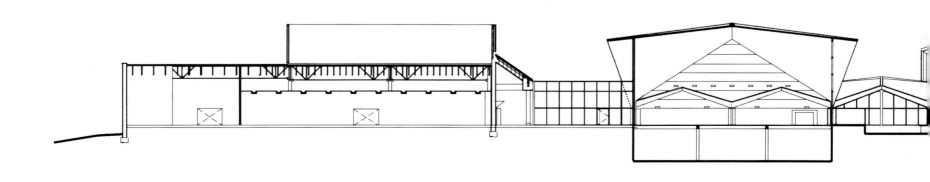

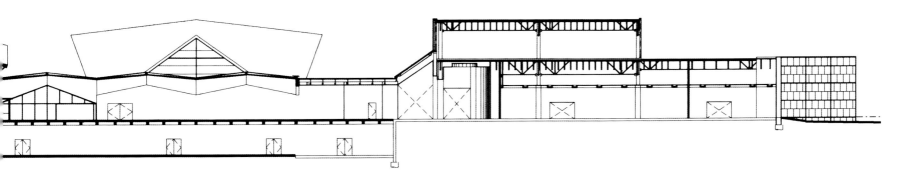

Long Sections

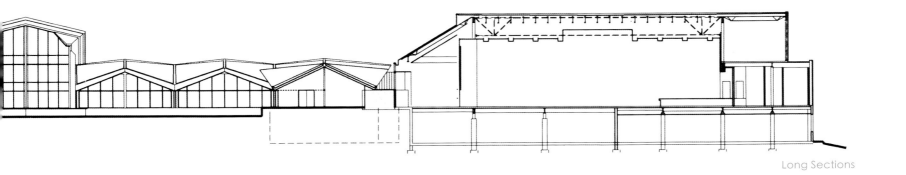

Long Sections

KEYWORD 关键词

- **Skin and Facade** 表皮与立面 | **Glass Facade** 玻璃立面
- **Materials** 材料 | **Glass** 玻璃
- **Ecological Building** 生态建筑

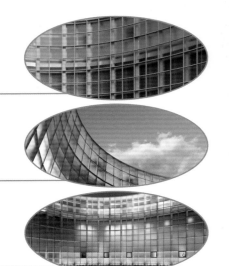

Location: Bellingham, Washington
Architect: Olson Kundig Architects
Size: 42,000 SF(3,901 sm)

项目地点：华盛顿州贝灵厄姆
建筑设计：Olson Kundig Architects
面积：3901 m²

Lightcatcher at the Whatcom Museum
沃特科姆博物馆莱特卡彻楼

Features 项目亮点

The most visible and innovative feature—the lightcatcher—a multi-functional translucent wall that reflects and transmits the Northwest's most precious and ephemeral natural resource, sunlight which features an interpretive exhibit about the low-impact development strategies.

建筑耀眼的、创新式的透明墙面，能够捕捉美国西北部最宝贵的自然资源——阳光，同时也更好的诠释了生态设计理念。

Overview 项目概况

The Lightcatcher at the Whatcom Museum is a LEED Silver regional art and children's museum.

It takes its name from its most visible and innovative feature—the lightcatcher—a multi-functional translucent wall that reflects and transmits the Northwest's most precious and ephemeral natural resource, sunlight. As architect Jim Olson describes it, "the lightcatcher wall celebrates the Northwest glass movement and glows like a yellowish agate from a nearby beach. I wanted to soften light like our clouds and create a sense of mystery like our mist and fog."

沃特科姆博物馆（Whatcom Museum）莱特卡彻楼（Lightcatcher Building）是华盛顿地区的一座获得LEED银级认证的博物馆。

莱特卡彻楼墙面高约11.3m，长约54.9m，是项目的核心部分。一座宽大的外部庭院，将博物馆的内外部空间连接起来。白天，墙面上的光孔把日光带入到大厅和展室，构成生态友好型的节能照明效果，同时也有助于建筑的通风。夜晚，墙面伴随着内部照明色彩的变化而微微发光，像一座灯笼一样带来温暖和吸引人的氛围。

The Lightcatcher Building at the Whatcom Museum

Section diagram with selected sustainable features

1. Green roof and impervious surfacing collect rainwater
2. Naturally ventilated Gallery / Circulation
3. Cistern to collect rainwater

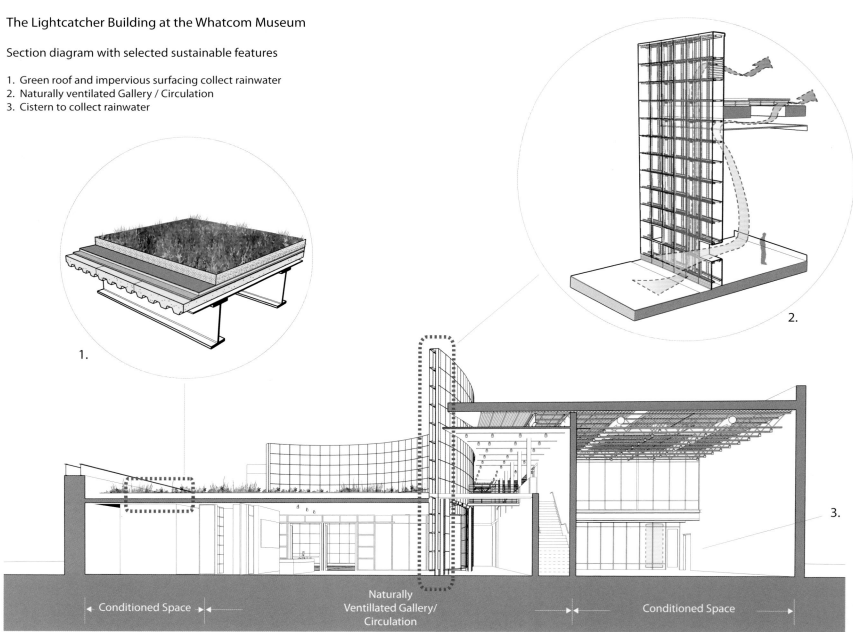

The Lightcatcher at the Whatcom Museum

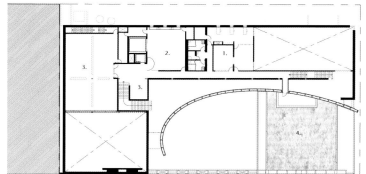

1. Administration
2. Education
3. Gallery
4. Green roof interpretive exhibit

Second Floor Plan

The Lightcatcher at the Whatcom Museum

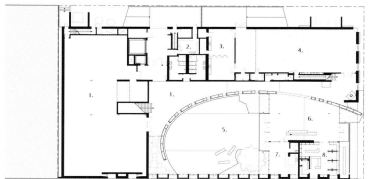

1. Gallery
2. Kitchen
3. Children's education
4. Family interactive gallery
5. Courtyard
6. Lobby
7. Cafe
8. Museum store

Main Floor Plan

Facade and Skin　立面与表皮

The lightcatcher, 37 feet high and 176feet long, is at the physical center of the project, gently curving to form a spacious exterior courtyard, while bridging the museum's interior and exterior spaces. During daylight hours, the light-porous wall floods the halls and galleries inside with a warm luminosity, serving as an elegant and energy-saving light fixture. The lightcatcher also helps ventilate the building. In the evening, the lightcatcher glows with the colors of the structure's interior illumination. Like a lantern, it provides a warm and welcoming beacon to the community.

这座建筑的名字来源于其最耀眼的、创新式的透明墙面，它能够捕捉美国西北部最宝贵的自然资源——阳光。就如它的设计者Jim Olson所描述的，莱特卡彻楼的墙面从附近的沙滩看过来就像是一块黄色的玛瑙，带一些雾气，显得飘渺而神秘。外墙和展馆的颜色很像是沙滩上的树皮和石块，而天花板则好似风化的浮木。

Structures and Materials　结构与材料

The first floor of the building features a lobby, three galleries, an interactive children's learning space, and other amenities. The building's second floor houses an additional exhibition gallery, meeting and classroom space, and museum offices. The single-story lobby is topped by a 3,000-square-foot green roof which features an interpretive exhibit about the roof and low-impact development strategies. The building utilizes natural materials endemic to the region.

莱特卡彻一楼设有一个大厅，三个画廊，一个儿童互动学习空间以及其他设施。二楼包括展览馆、会议室、教室、以及博物馆办公室。大堂屋顶是一个287.7 m^2的绿色空间，用来展示和解释环保理念。大楼建筑材料是当地最常见的。

KEYWORD 关键词	Skin and Facade 表皮与立面	Concrete Facade 混凝土立面
	Materials 材料	Preset Concrete Slab 预制混凝土板
	Details 细部	

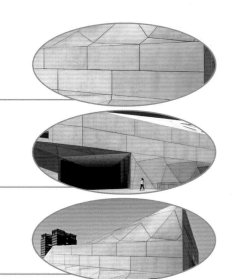

Location: Tel Aviv, Israel
Developer: Motti Omer, Director and Chief Curator
Architectural Design: Preston Scott Cohen
Material: precast reinforced concrete

项目地点：以色列特拉维夫
开发商：Motti Omer（馆长）
建筑设计：Preston Scott Cohen
材质：预制钢筋混凝土

Tel Aviv Museum of Art
特拉维夫艺术博物馆

Features 项目亮点

A spiraling, top-lit atrium, whose form is defined by subtly twisting surfaces that curve and veer up and down through the building, serve as the surprising, continually unfolding vertical circulation system; while the natural light from above is refracted into the deepest recesses of the half-buried building.

建筑内有一个成螺旋状上升的明亮中庭，经过巧妙设计，形成惊人的垂直环流交通，并使得顶部光线直达建筑底层最深处。

Overview 项目概况

Design and construction of a freestanding new building for the complex of the Tel Aviv Museum of Art, the leading museum of modern and contemporary art in Israel, housing an installation of the Museum's comprehensive collection of Israeli art, as well as its architecture and design galleries, drawings and prints galleries, photography study center, art library, new auditorium, a large gallery for temporary exhibitions and public amenities, the Herta and Paul Amir Building is intended to create an outstanding, forward-looking work of architecture for the Municipality of Tel Aviv.

Location The Museum is located in the heart of Tel Aviv at 27 Shaul Hamelech Boulevard, set back from the street behind a large plaza. The Ministry of Justice stands to the east; the Beit Ariela Municipal Library and the Center for the Performing Arts are to the west. The site for the Amir Building is a triangular plot between the existing Museum complex, the Library and the Center for the Performing Arts.

特拉维夫艺术博物馆无疑是以色列领先的现代和当代艺术博物馆建筑，建筑全方位的收集以色列的艺术品，备有画廊、研究中心、图书馆、礼堂、临时画廊以及其它公共设施。这是一个具有前瞻性的建筑。

建筑坐落于特拉维夫 Shaul Hamelech 大道 27 号，背临一个大广场，位于法院东侧，图书馆和剧院的西侧。建筑在这三个建筑之间，有着复杂微妙的关联。

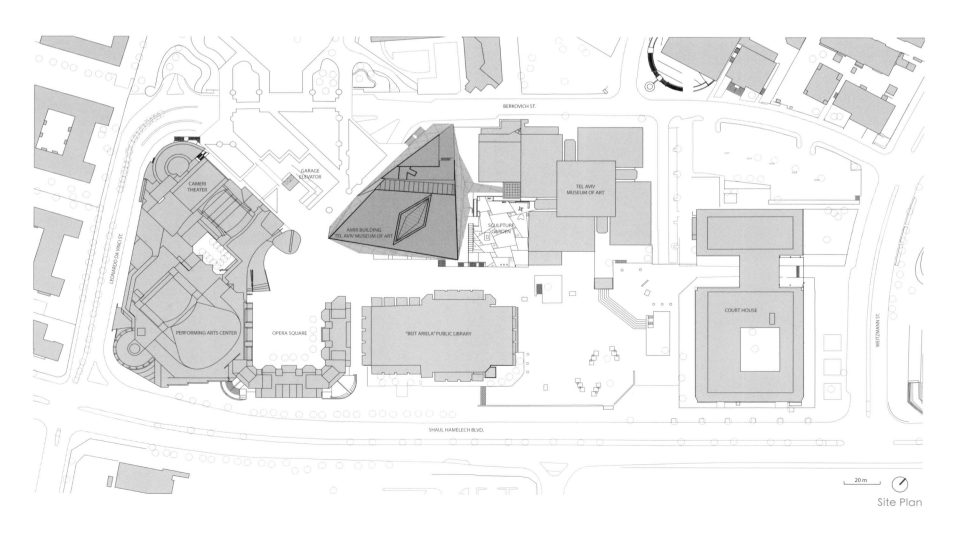

Site Plan

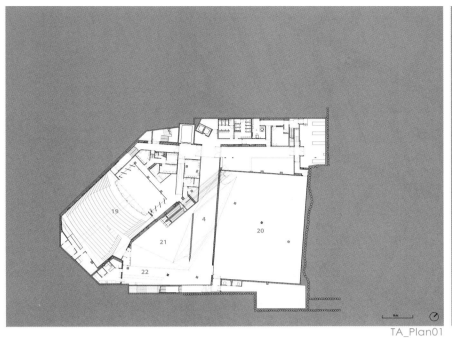

TA_Plan01

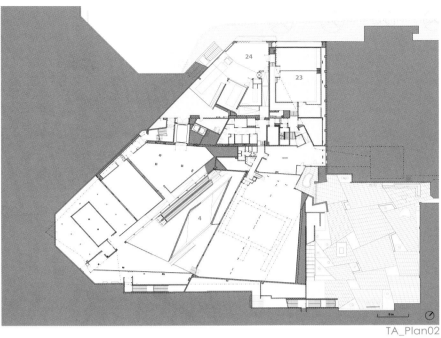

TA_Plan02

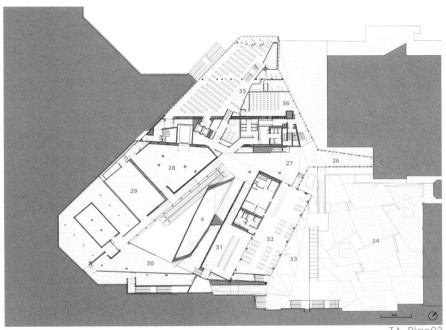

TA_Plan03

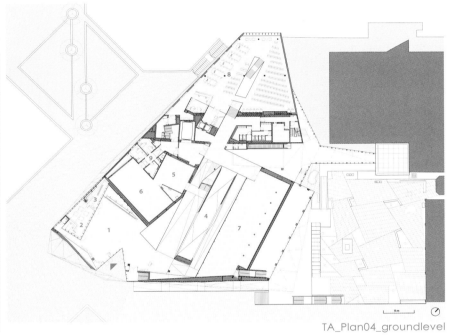

TA_Plan04_groundlevel

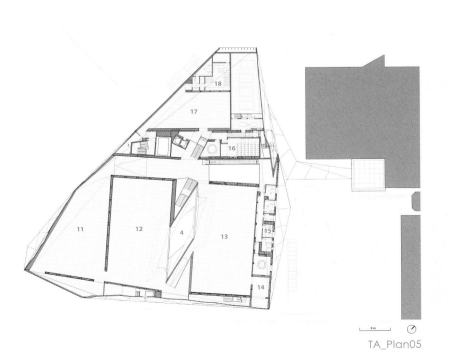

TA_Plan05

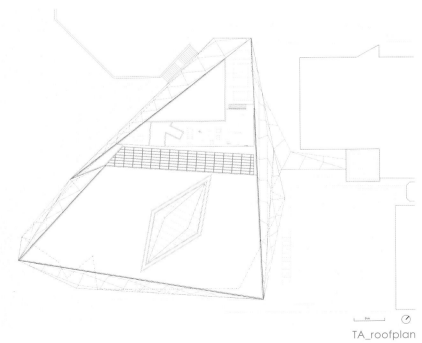

TA_roofplan

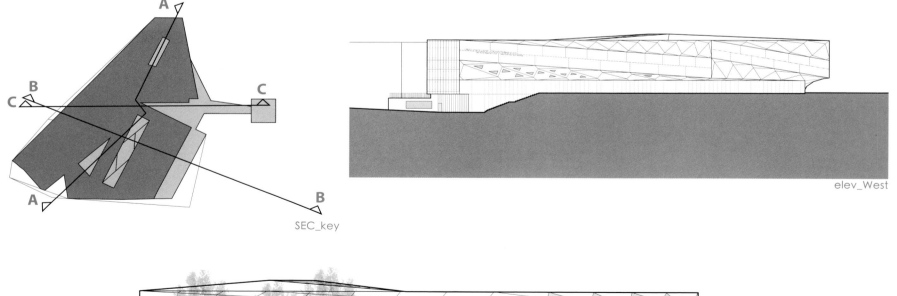

SEC_key

elev_West

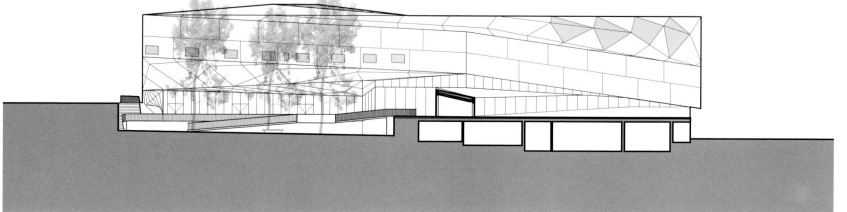

elev_East

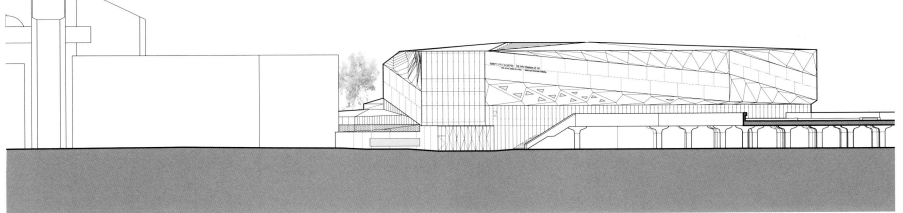

elev_North

Shape and Facade 造型与立面

The design for the Amir Building arises directly from the challenge of providing several floors of large, neutral, rectangular galleries within a tight, idiosyncratic, triangular site. The solution is to "square the triangle" by constructing the levels on different axes, which deviate significantly from floor to floor. In essence, the building's levels—three above grade and two below—are structurally independent plans stacked one on top of the other.

三角形的场地现状使得建筑内部画廊成长条形并排列紧凑，同时带来楼层设计的挑战。设计师将楼层设计为错层，因此，建筑的三层中有两层的顶部结构还能提供额外的独立层面。

Materials and Forms　材料与形式

In this way, the Amir Buliding combines two seemingly irreconcilable paradigms of the contemporary art museum: the museum of neutral white boxes, which provides optimal, flexible space for the exhibition of art, and the museum of spectacle, which moves visitors and offers a remarkable social experience. The Amir Building's synthesis of radical and conventional geometries produces a new type of museum experience, one that is as rooted in the Baroque as it is in the Modern.

These levels are unified by the "Lightfall": an 87-foot-high, spiraling, top-lit atrium, whose form is defined by subtly twisting surfaces that curve and veer up and down through the building. The complex geometry of the Lightfall's surfaces (hyperbolic parabolas) connect the disparate angles of the galleries; the stairs and ramped promenades along them serve as the surprising, continually unfolding vertical circulation system; while the natural light from above is refracted into the deepest recesses of the half-buried building. Cantilevers accommodate the discrepancies between plans and provide overhangs at the perimeter.Conceptually, the Amir Building is related to the Museum's Brutalist main building (completed 1971; Dan Eytan, architect). At the same time; it also relates to the larger tradition of Modern architecture in Tel Aviv, as seen in the multiple vocabularies of Mendelsohn, the Bauhaus and the White City.The gleaming white parabolas of the façade are composed of 465 differently shaped flat panels made of pre-cast reinforced concrete. Achieving a combination of form and material that is unprecedented in the city, the façade translates Tel Aviv's existing Modernism into a contemporary and progressive architectural language.

该建筑结合了两种不同的现代艺术博物馆典范：将中性的、最适合展览的、具有灵活空间的展厅与奇妙的博物馆动线体验结合起来，形成非凡的情景体验。先锋与传统的在这里碰撞，产生了一个全新的博物馆。

建筑有一个26.5 m高，成螺旋状上升的明亮中庭。沿中庭周边的形式经过巧妙的定义和设计，复杂的双曲线多角度楼梯串联起展厅，形成惊人的垂直环流交通，并使得顶部光线直达建筑底层最深处。从概念上讲，该博物馆脉承野兽派主要代表作(1971; Dan Eytan, architect)，同时吸取现代建筑精华。465个预制混凝土板形成不同形状的表面。这是一个将形式和材料充分落实的，根植于特拉维夫的现代主义建筑。

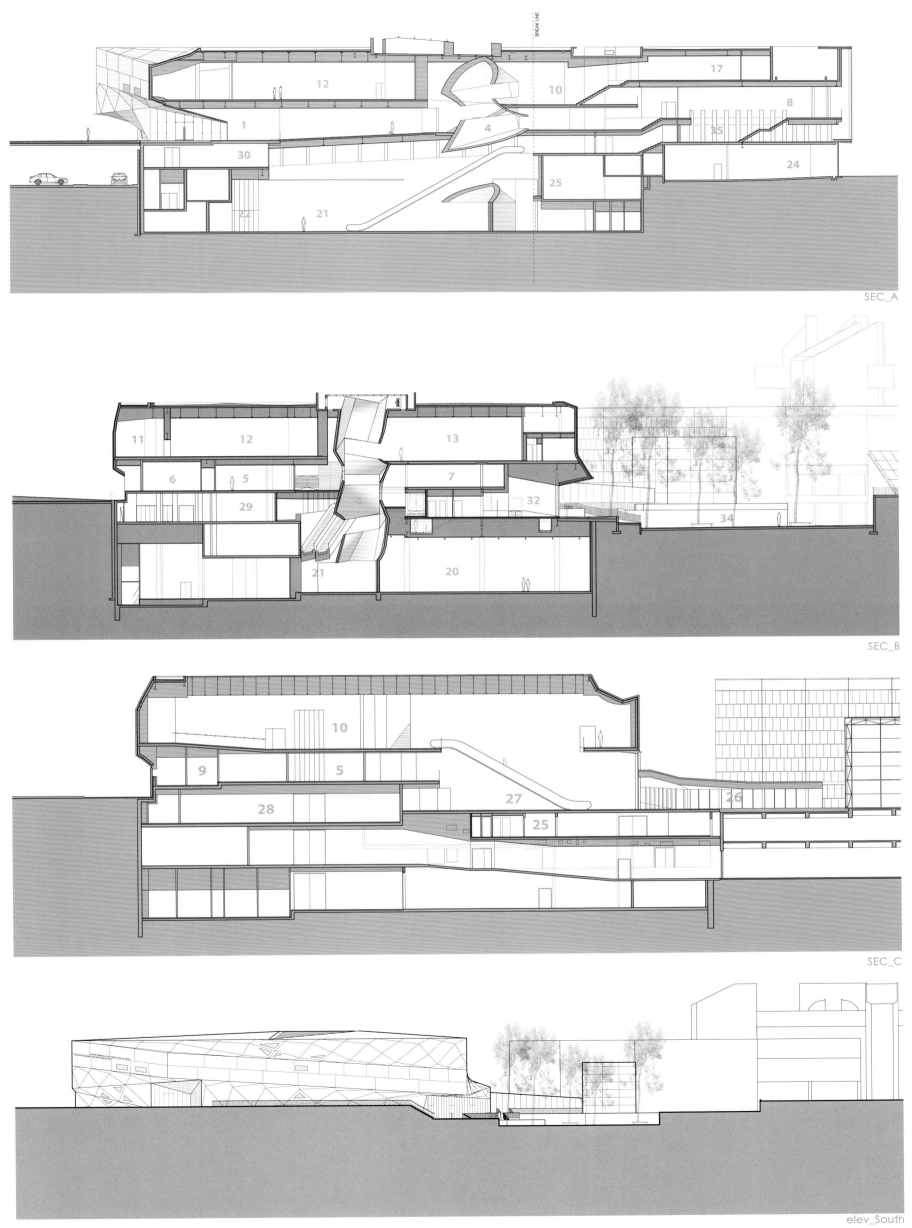

KEYWORD 关键词	Skin and Facade 表皮与立面	Red Alucobond Curvy Slats, Metallic Facade 红色闪银色曲线板条，金属立面
	Materials 材料	Composite Panel 复合板
	Detail 细部	

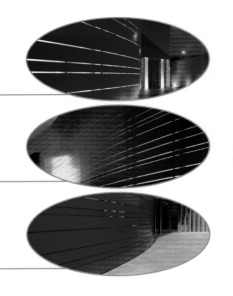

Location: Milan, Italy
Client: Gruppo Cimbali
Architects: Arkispazio
Head designers: Paolo Balzanelli, Valerio Cometti
Structural Engineering: Francesco Terreni
Built Surface: 1,800 m²
Photography: Arkispazio

项目地点：意大利米兰
客户：Gruppo Cimbali
建筑设计：意大利 Arkispazio 建筑事务所
主设计师：Paolo Balzanelli, Valerio Cometti
结构工程：Francesco Terreni
建成面积：1800 m²
摄影：Arkispazio

Museum MUMAC Museum of Coffee Machine
MUMAC 咖啡机博物馆

Features 项目亮点

The facades of the museum are covered with strips of metal "red Cimbali", sinuous and enveloping to resemble the waves of hot coffee, which at night filters the artificial light creating a striking illuminated reticle that evokes the energy of MuMAC.

建筑外墙选用红色的带状装饰，象征热咖啡飘动的热气，加上背景光的照射，使其散发出引人注目的光芒。

Overview 项目概况

The MuMAC, Museum of Coffee Machine, was designed by architect Paolo Balzanelli owner of Arkispazio and engineer Valerio Cometti to celebrate the centenary of the Cimbali Group. The Mumac lies in the establishment of the Cimbali Group in Milan, inside the building previously used as a warehouse, within which are located the museum area, an area for temporary exhibitions and dedicated to the culture of coffee.

For visitors to the Mumac, there was created a new entrance through which you can access an area bounded by a coffee-colored fence. The back wall of the fence is marked by nine trees which divide it into 10 equal spaces: ten decades of the century that symbolize life and achievements of Gruppo Cimbali. The opposite side of the building is facing industrial area of the property provides entrance for corporate visitors. The museum area offers an exposition divided into six historical periods from the beginning of the century to the present day: The early years, The age of Rationalism, Invention of the lever, Under the banners of design, The International dimension and The New Millennium.

MUMAC 咖啡机博物馆的设计工作由意大利 Arkispazio 建筑师事务所的建筑师 Paolo Balzanelli 和工程师 Valerio Cometti 合作完成。它位于米兰的 Cimbali 集团的仓库区域内，作为一个临时的咖啡文化展厅。

展厅内部根据不同的年代将展区划分成不同的部分，方便游客参观。建筑前面用 9 棵树分成 10 等分，以此来展示一个世纪以来 Cimbali 集团形象的变化，同样内部则分为六个区域（The early years, The age of Rationalism, Invention of the lever, Under the banners of design, The International dimension and The New Millennium），通过他们的咖啡机来展示咖啡的发展史和企业文化。

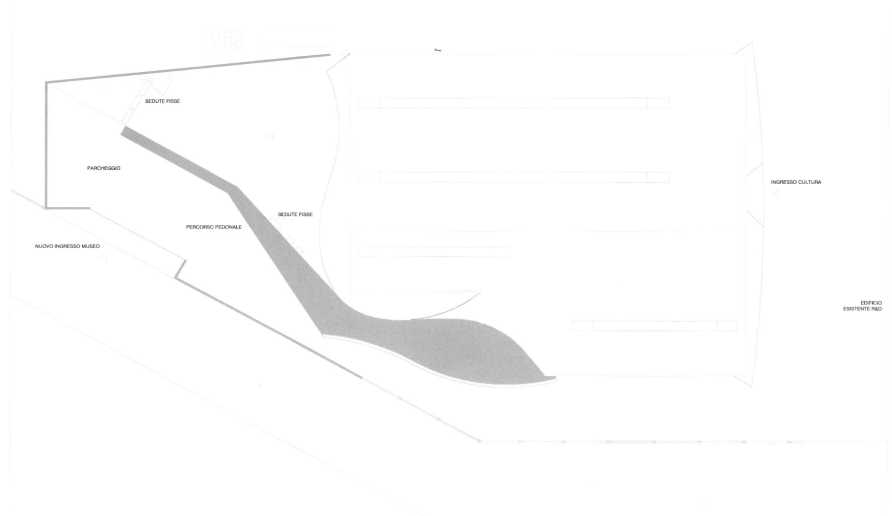

PLANIMETRIA AREA ESTERNA

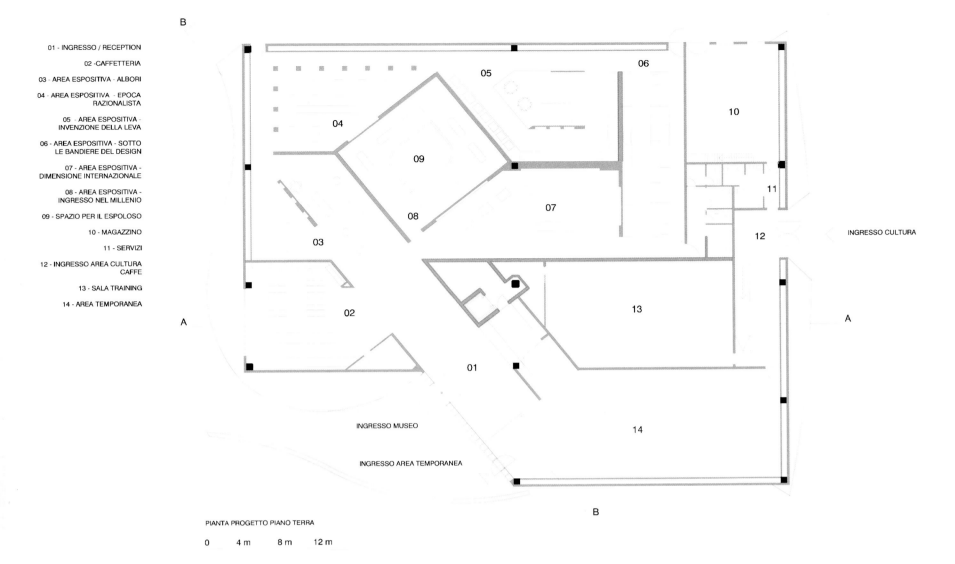

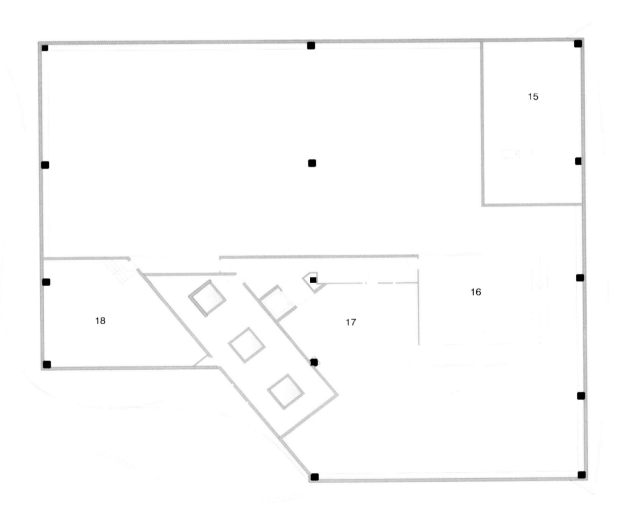

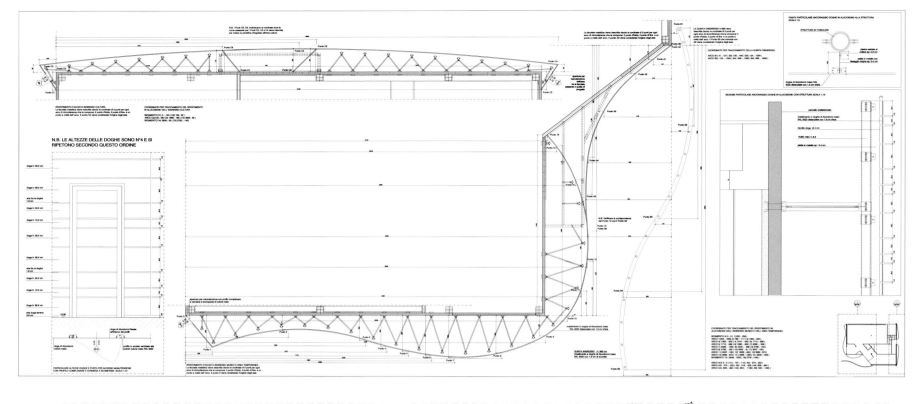
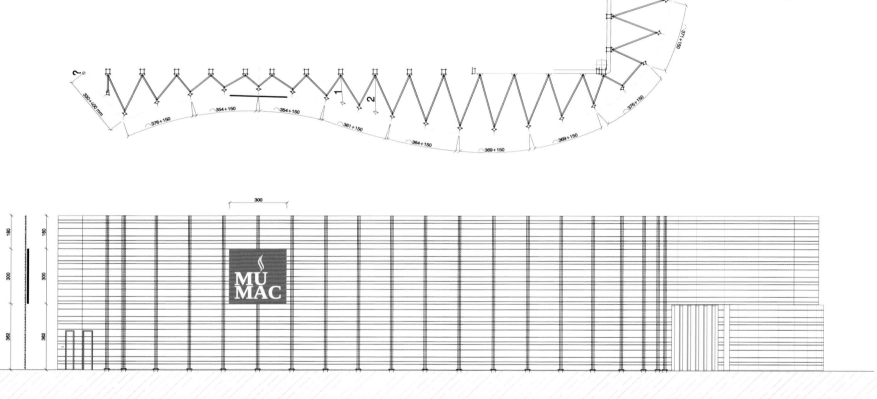

PROSPETTO NORD

PROSPETTO OVEST

PROSPETTO EST

PROSPETTO SUD

0   4 m   8 m   12 m

Skin and Materials   表皮与材料

MuMAC tells 100 years of the history of objects that find home within its 1,800 square meters, besides museum there is an area dedicated to the "world of coffee" and an area for temporary exhibitions. The facades of the museum are covered with strips of metal "red Cimbali", sinuous and enveloping to resemble the waves of hot coffee, which at night filters the artificial light creating a striking illuminated reticle that evokes the energy of MuMAC.

Material the Arkispazio use is reyonbond by Alcoa, that is the same metal composite as alucobond, they are both "a composite panel consisting of an extruded thermoplastic compound core that's fusion bonded between two sheets of coil-coated aluminum. The result is a highly corrosion-resistant, rigid-yet-flexible material that weighs 3.4 times less than steel and 1.6 times less than pure aluminum. Its extreme formability makes it ideal for curves and unique design accents." Façade of the museum is made of curvy slats of a four different heights, with 1.8 cm gaps between each one to let the light pass through. They were assembled by Metalcop Srl.

博物馆占地 1 800 m²，展示着咖啡文化的百年历史。建筑外墙选用红色的带状装饰，象征热咖啡飘动的热气，背景光的照射，让 Mumac 的标识更醒目。到晚上，人工灯光可以使其散发出引人注目的光芒。

项目采用两种材料 reyonbond 和 alucobond，它们都是由铝镀薄板压制热塑性材料而成的复合面板，具有高防腐性，刚柔并济，比纯钢和纯铝要轻的多，再加上极佳的成型性能，无疑是该曲线立面的首选材料。立面的弯曲面板，以四种不同的高度呈现，每个高度有异的连接处都有 1.8 cm 的空隙，引入阳光。

KEYWORD 关键词	Skin and Facade 表皮与立面	Wooden Facade 木制立面
	Materials 材料	Spruce Wood Planks, Alder Planks and Nanotech Wax 云杉板、桤木板、纳米蜡
	Shape 造型	Curved Surface 曲面造型

Location: Helsinki, Finland
Client: Helsinki Parish Union and the City of Helsinki
Architects: K2S Architects
Total Floor Area: 352 m²
Photography: Tuomas Uusheimo, Marko Huttunen

项目地点：芬兰首都赫尔辛基
客户：Helsinki Parish Union and the City of Helsinki
建筑事务所：芬兰 K2S Architects 建筑事务所
总建筑面积：352 m²
摄影：Tuomas Uusheimo, Marko Huttunen

Kamppi Chapel
康比教堂

Features 项目亮点

The project widely uses wooden materials in the surface, interior walls and furniture, etc., to make a warm and quiet environment for the chapel.

项目大量采用木料，用于外立面、室内墙面、家具等，为教堂营造温暖安静的氛围。

Overview 项目概况

The Kamppi Chapel is located on the south side of the busy Narinkka square in central Helsinki. It offers a place to quiet down and compose oneself in one of Finland's most lively urban spaces. With its curved wood facade, the small sacral building flows into the city landscape. Simultaneously the chapel gently shaped interior space embraces visitors and shields them from the bustling city life outside.

康比教堂位于赫尔辛基中心区熙熙攘攘的纳瑞卡广场南侧。在芬兰最活跃的城市空间之中，小教堂提供了一个能让人静下来思考自我的地方。通过弯曲的木制立面，这座小型的圣堂建筑融入到城市景观里。与此同时，小教堂柔和造型的室内环境，像盾牌似的将到访者与外面庸碌的城市生活隔开。

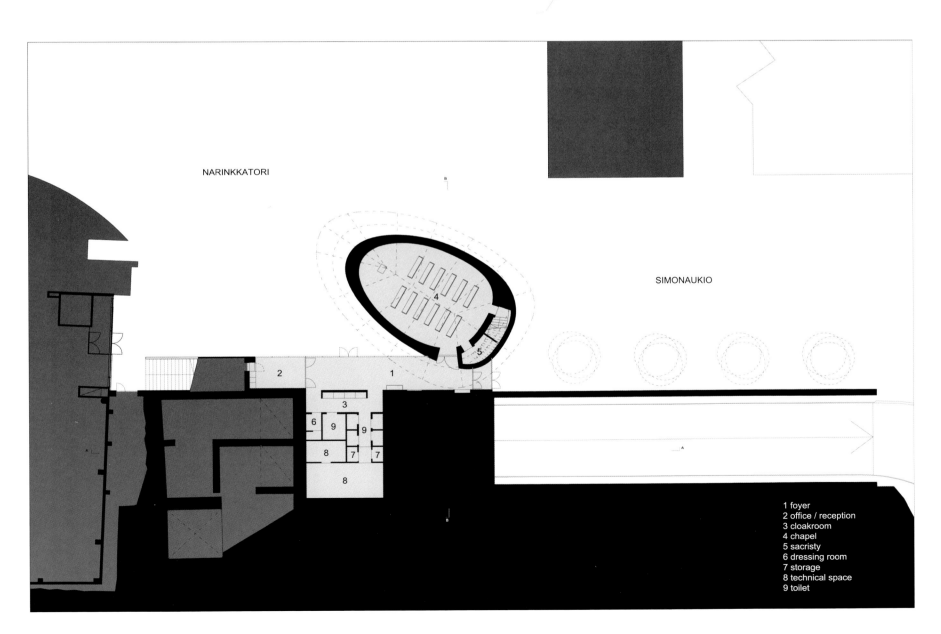

1 foyer
2 office / reception
3 cloakroom
4 chapel
5 sacristy
6 dressing room
7 storage
8 technical space
9 toilet

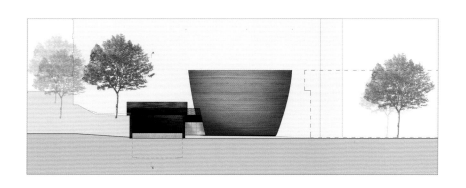
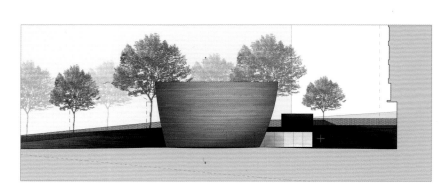
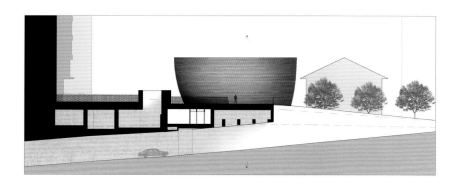
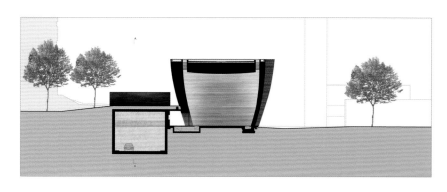

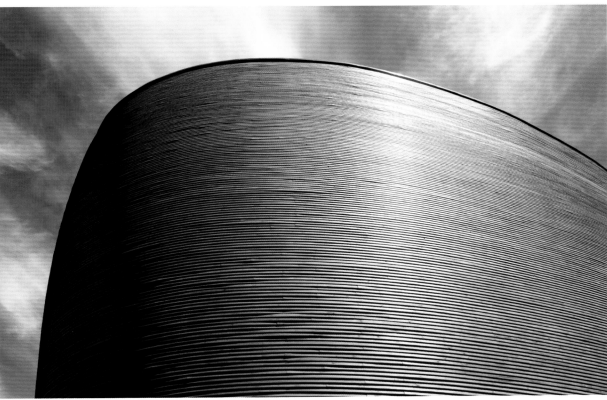

Surface and Facade 表皮与立面

The facades are made of sawn-to-order horizontal finger jointed spruce wood planks, which are treated with a pigmented transparent nanotech wax. The constructive frame consists of cnc-cut gluelam elements.

外立面由预先锯好的云杉板水平连接而成，表面上有透明染色的纳米蜡。结构框架也结合了数控技术。

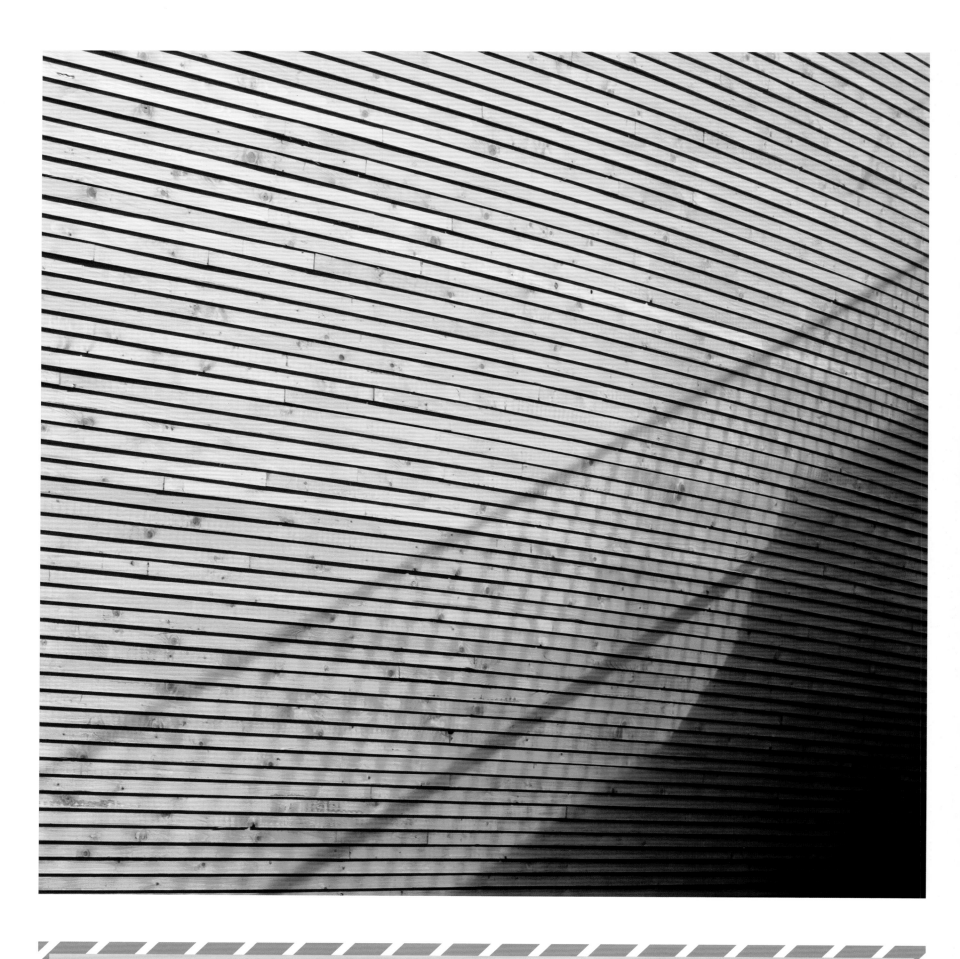

Structure and Materials 结构与材料

The chapel can be approached from all directions. From the direction of the Simonkatu, one arrives at a small square opening up towards the Narinkka square. From there, a flight of stairs leads down to the entrance level. Entrances are located in two glass facades facing the Narinkka square and the Lasipalatsi building. Only the actual chapel space is located in the wooden volume. Secondary spaces are located in a space opening up towards the square. The entrance space doubles as exhibition space, in which one also encounters clergymen and social workers.

The sacral space is a calm space, in which the lively neighborhood seems distant. Light touching down on the curved surface and the feeling of warm materials define the space. The chapel's inner walls are made of thick oiled alder planks. The furniture is also made of solid wood.

教堂能够从四面八方进入。在斯蒙卡图的方向，有一个朝着纳瑞卡广场的小广场。从那里，沿着一段楼梯向下，就可以走到入口处。入口分别位于面对纳瑞卡广场和拉斯帕拉茨大楼的两个玻璃立面上。在木造空间内，才是真正的教堂空间。次一级的空间位于对着广场的开放空间里。入口空间是展示空间的两倍大，其中一个提供给神职人员与社会工作者使用。

圣堂空间是非常安静的空间，与它活跃的周边区域大相径庭。轻轻触摸曲线的表面，感受一下这个由温暖材料构筑的空间。教室内墙由上过油的厚重桤木板制成。家具也是实木制的。

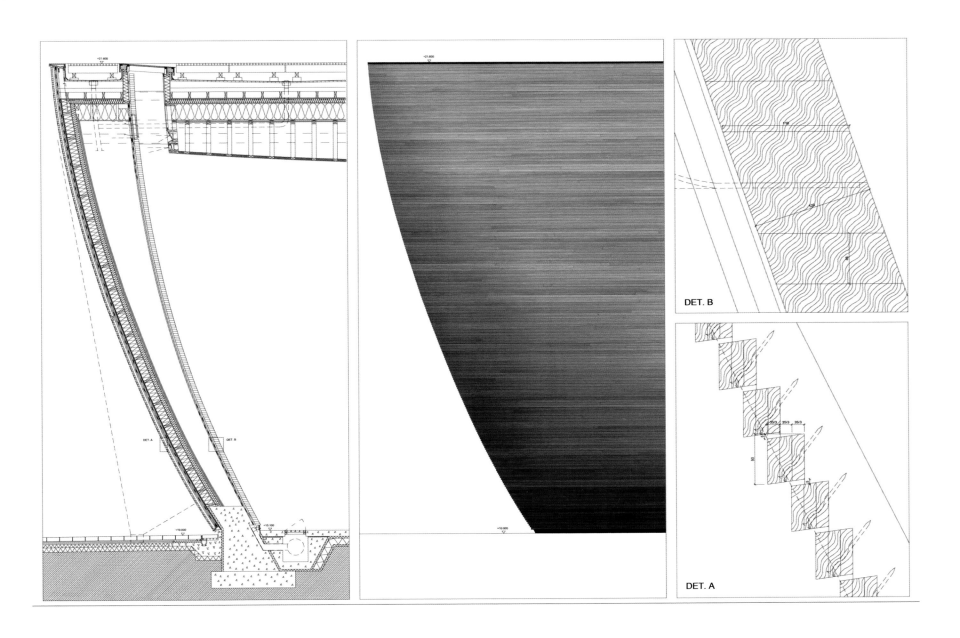
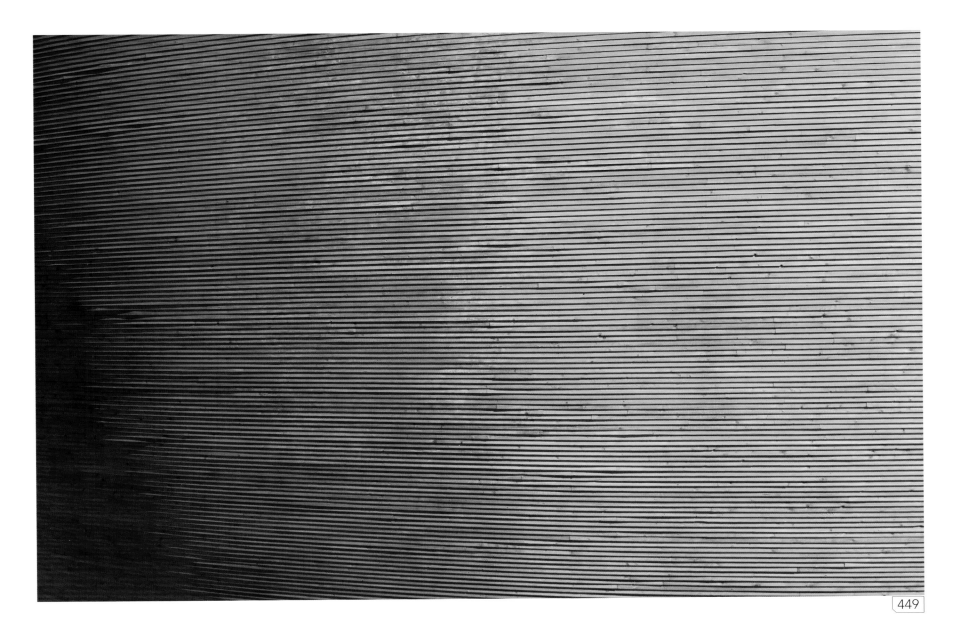

KEYWORD 关键词		
Skin and Facade 表皮与立面	Stone Facade 石材立面	
Materials 材料	Honeycomb Panels, Stone 蜂窝板，石板	
	Spiral Structure 螺旋形式	

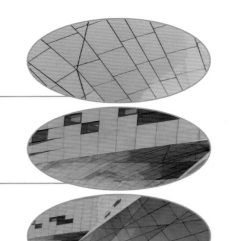

Location: Taiyuan, China
Client: Taiyuan City GovernmentTotal
Architectural Design: Preston Scott Cohen
Area: 32,500 m²

项目地点：中国山西省太原市
客户：太原市政府
建筑设计：普雷斯顿·斯科特·科恩
面积：32 500 m²

Taiyuan Museum of Art
太原美术馆

Features 项目亮点

Exterior light weight honeycomb panels with stone veneer produce an evocative and elusive material effect and the perception of an exceptional scale. The panels are reflective as if metallic.

外立面上的轻型蜂窝板贴有薄片石材镶面，如金属一般可以反光，产生出一种引人深思而又难以捉摸的材料效果。

Overview 项目概况

The Taiyuan Museum of Art works as a cluster of buildings unified by continuous and discontinuous promenades both inside and outside. The building responds to the urban parkscape in which it is set; visitors are encouraged to pass through the building while not entering into the museum itself. An exterior ramp threading through the building connects the heterogeneous hardscapes, lawns and sculpture gardens. The integration of building and landscape registers multiple scales of territory ranging from the enormity of the adjacent Fen River to the intimacy of the museum's own particular spatial episodes.

Inside, the security of museum space is maintained by a highly controlled interface between gallery and non-gallery programs including an auditorium, bookstore, restaurant, library, education center, and administrative wing. The individual sets of elevators and cores are distributed to guarantee easy access and easy divisibility between zones regulated by different schedules and rules of access. At the garage level, the services are intricately planned in order not to interfere with parking lots for staff and public.

The museum galleries are organized to ensure maximum curatorial flexibility. The galleries can be organized into a single, spiraling sequence for large chronological exhibitions or into autonomous clusters operating independently. For visitors architectural cues offer - the placement of ramps and portals, the expansion and contraction of space - provide a means of wayfinding. The building gives visitors the freedom either to follow a predetermined chronological sequence or to skip from one set of galleries to another, in a nonlinear fashion.

　　这座美术馆由一系列连续与不连续的空间构成，在给游赏者一个连续空间序列印象的同时又允许他们采取非线性的方式略去其中一些展区，自主地选择参观路线。一条自室外开始的坡道在建筑内外纵横连接了诸如硬地广场、草坪以及雕塑花园等多个空间。建筑与景观的有机融合使得美术馆得以在应对汾河风光带大格局的同时确保自身空间序列的完整性与特殊性。

　　博物馆内部的安全性能由画廊与非画廊空间之间控制程度极高的交汇处保证，其中非画廊空间包括礼堂、书店、餐厅、图书馆、教育中心以及行政侧楼。电梯和交通核心筒分开设置，确保通行便利，同时容易区分使用时间段与通道规定各异的区域。在车库那一层，为了不干扰职工和公众停车场，建筑设备设计得比较复杂。

　　博物馆画廊的设计以尽量确保策展工作的灵活性为原则。画廊可为按时间顺序排列的大型展览设计为螺旋形式，也可设计为单独工作的互不干扰的空间。坡道和入口的安置、空间的膨胀与收缩这些建筑线索都为游人们提供了一种指路的手段。建筑予以游人充分自由，要么遵循预定的时间序列，要么不走寻常路，从一组画廊直接穿到另一组。

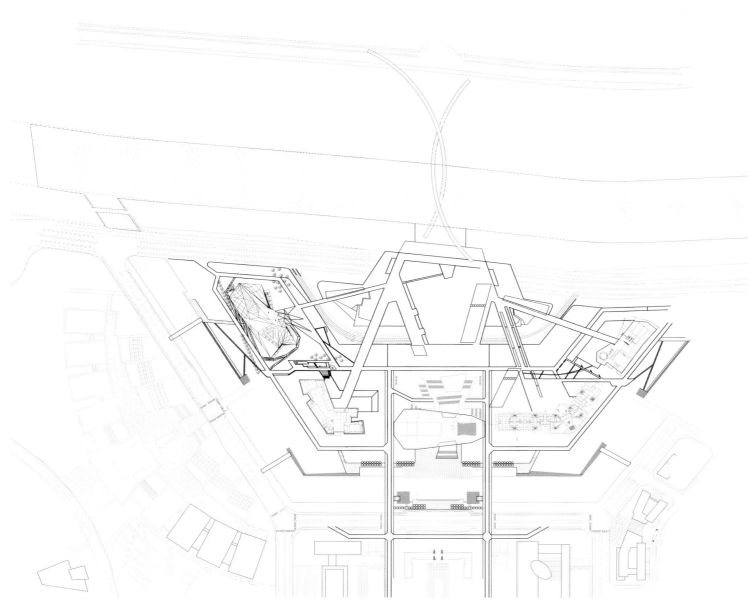

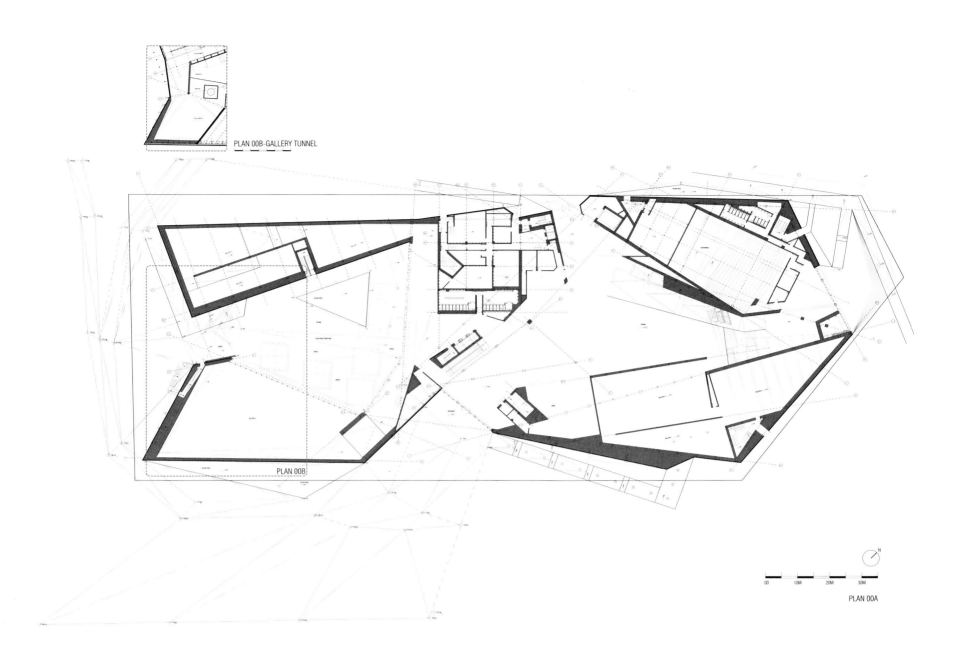

PLAN 00B-GALLERY TUNNEL

PLAN 00A

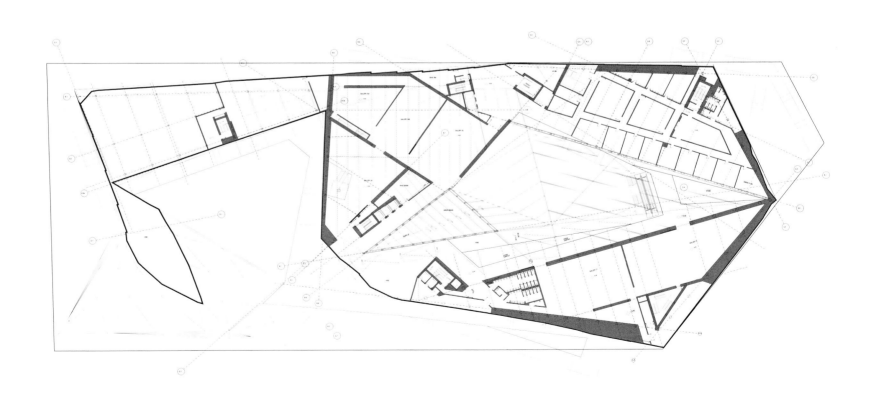

PLAN 03

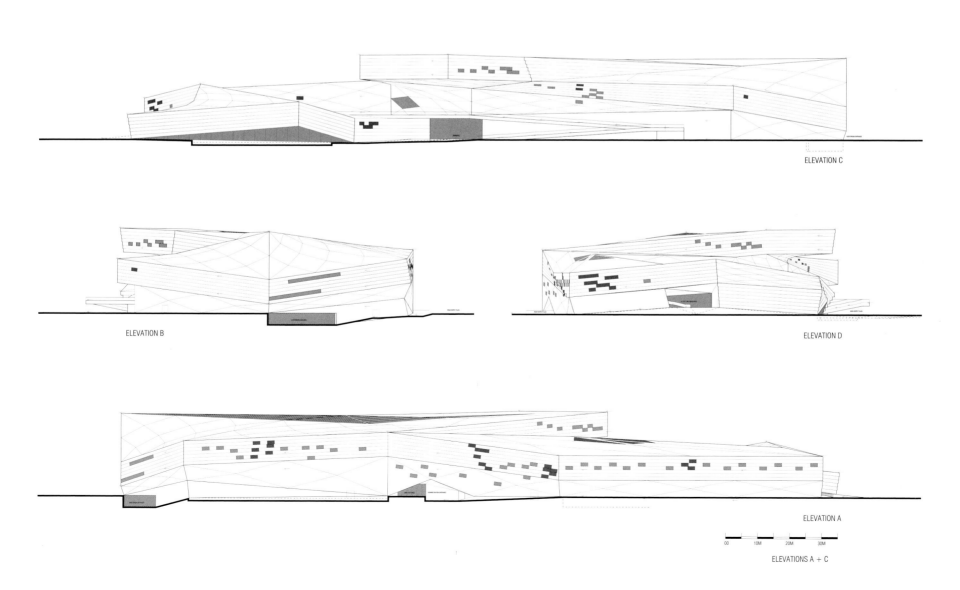

ELEVATION C

ELEVATION B

ELEVATION D

ELEVATION A

ELEVATIONS A + C

SECTION A-A   SECTION B-B

Skin and Structure   表皮与结构

Exterior light weight honeycomb panels with stone veneer produce an evocative and elusive material effect and the perception of an exceptional scale. The panels are reflective as if metallic, seemingly too large to be stone panels, but clearly possessing the properties of both materials. Advanced parametric software allowed panels to conform to standard widths, reducing material waste.

室外轻型蜂窝板贴有薄片石材镶面，产生出一种引人深思而又难以捉摸的材料效果，令人感受到极为特殊的建筑规模。板材如金属一般可以反光，看上去特别大，都不像石板，但显然拥有两者的材料特性。先进的参数化软件可按标准宽度制造板材，减少了材料的浪费。

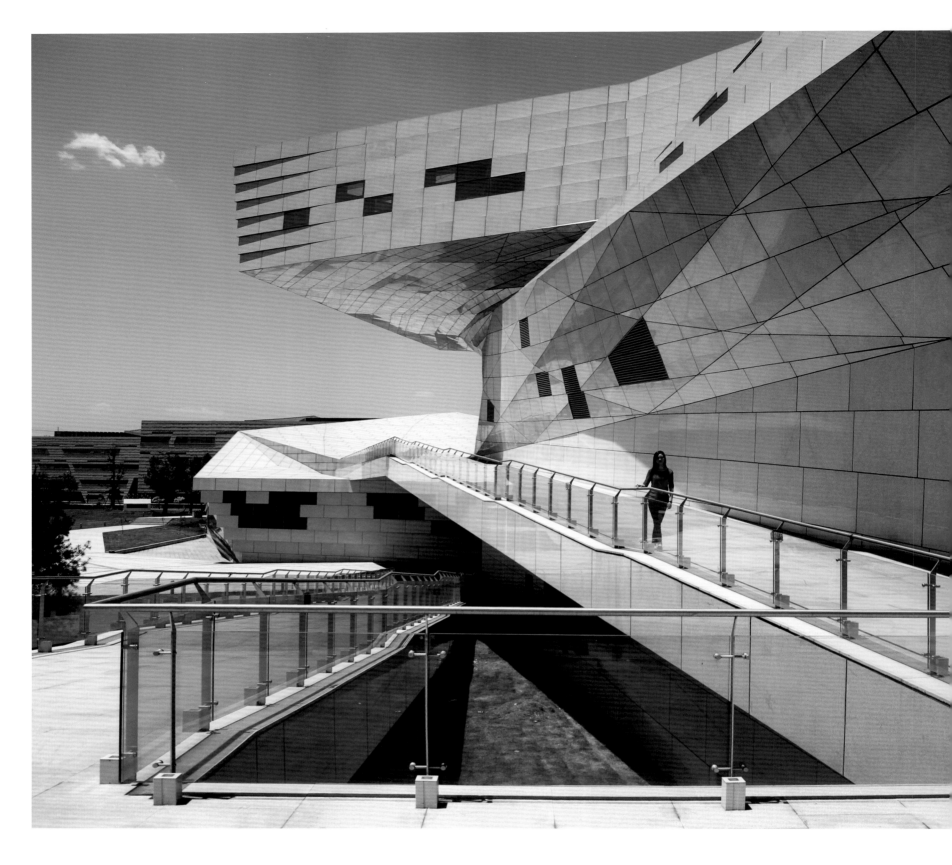

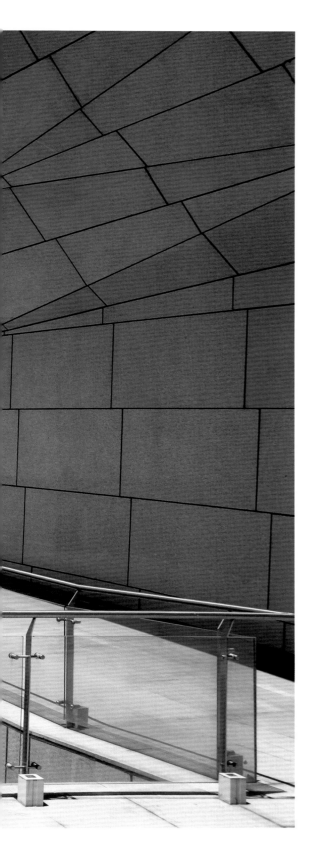
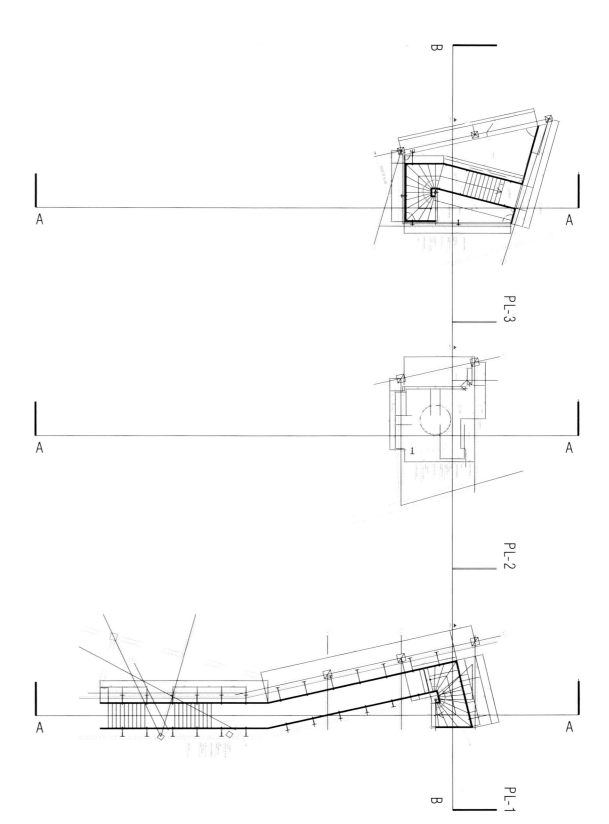

KEYWORD 关键词	Skin and Facade 表皮与立面	Metal Facade 金属立面
	Materials 材料	Glass 玻璃
	Details 细部节点	

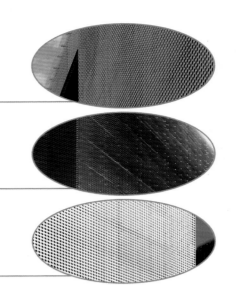

Location: Tampa, Florida, USA
Construction area: 6131.6 m²
Designer: Stanley Saitowitz (SSNA Architects)
Design team: Neil Kaye, Markus Bischoff
Lighting: Jason Edling, Key Riley Anderson
Photographer: Richard Barnes, James Ostrand

项目地点：美国佛罗里达州坦帕
建筑面积：6131.6 m²
设计公司：Stanley Saitowitz（SSNA 建筑事务所）
设计团队：Neil Kaye, Markus Bischoff
灯光设计：Jason Edling, Key Riley Anderson
摄影师：Richard Barnes, James Ostrand

Tampa Museum of Art Tampa, Florida
佛罗里达坦帕美术馆

Features 项目亮点

A glass pedestal supports the jewelbox of art above. The building floats in the park, embracing it with its overhanging shelter and reflective walls. The building is not only in the landscape, but is the landscape.

建筑底部的玻璃基础使建筑整体像一个漂浮于公园之上的珠宝盒，不但很好地融入周边风景，并且创造出更为绚丽迷人的风景。

Overview 项目概况

Museums began in ancient times as Temples, dedicated to the muses, where the privileged went to be amused, to witness beauty, and to learn. After the Renaissance museums went public with palatial structures where the idea of the gallery arose, a space to display paintings and sculpture. Later, museums became centers of education, researching, collecting, and actively provoking thought and the exchange of ideas. By presenting the highest achievements of culture, museums became a stabilizing and regenerative force, crusading for quality and excellence. The role of the modern museum is both aesthetic and didactic, both Temple and Forum.

The design of contemporary museum can be characterized by two polar approaches. On the one-hand buildings which aim to be works of art in themselves, independent sculptural objects as signatures of their architects. On the opposite end of the spectrum are museums as containers, as beautiful jewel boxes, treasure chests whose sole purpose is to be filled with art, like the Tampa Museum.

该博物馆最初的形式是供奉缪斯的古代寺庙，特权阶级可在这里欣赏精致的艺术品并向它们学习。文艺复兴之后，博物馆建筑越发雄伟、壮丽，并开始向市民公开，通过建筑内廊和展览空间向人们展示绘画和雕塑艺术品。随后博物馆逐渐成为培养艺术素养，研究、收集艺术品以及人们思考与交换意见的中心。通过呈现高品质的艺术精品，博物馆正在稳步地进行变革。现代博物馆应当兼具形式美和文化传承，融会殿堂与论坛的双重气质。

当代博物馆的设计可以具有两种截然不同的特征，一方面建筑师可以将建筑作为一个独立的艺术品用以彰显自己的艺术个性，另一方面将博物馆看作一个容器——一个美丽的珠宝盒——她存在的唯一目的就是装满艺术瑰宝，这正是坦帕艺术博物馆的创作初衷。

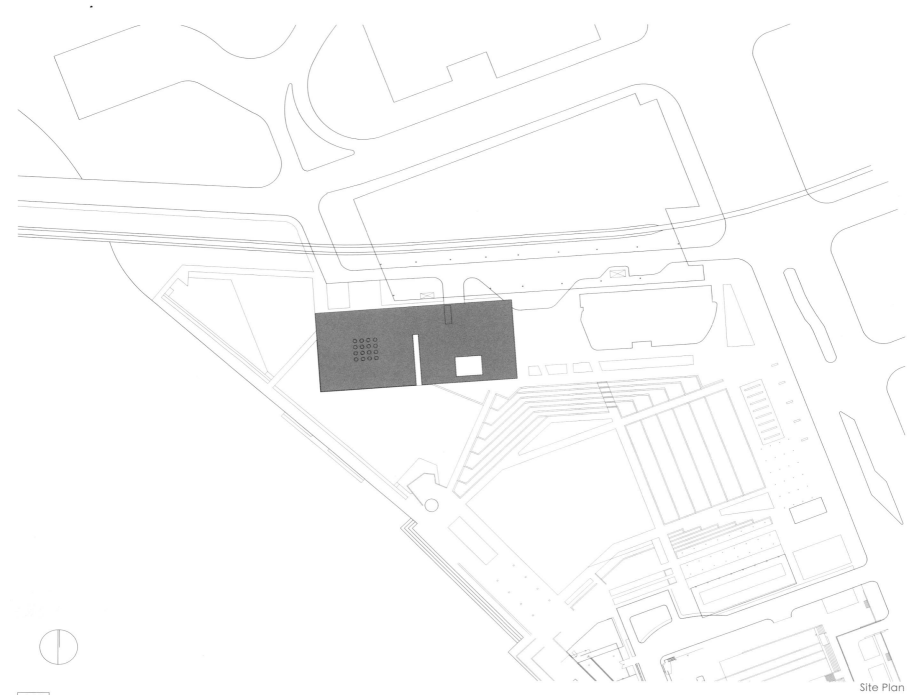

Site Plan

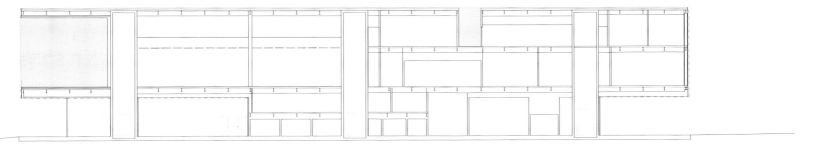

longitudonal sections

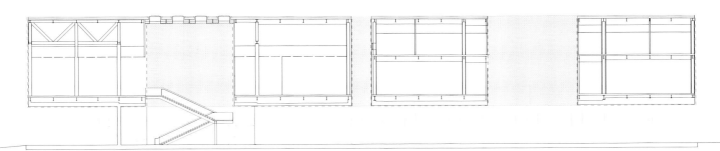

longitudonal sections

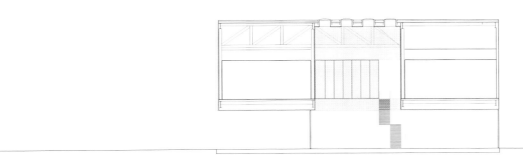

cross sections

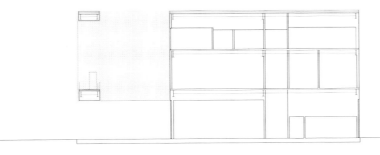

cross sections

Structures and Material　结构与材料

This museum is a neutral frame for the display of art, an empty canvass to be filled with paintings. It is a beautiful but blank container, a scaffold, to be completed by its contents. We are interested in openness, in unknown possibilities in the future, in Architecture as infrastructure. We have created compelling space in the most discreet way, avoiding the building as an independent sculptural object, and using space and light to produce form.

A glass pedestal supports the jewelbox of art above. The building floats in the park, embracing it with its overhanging shelter and reflective walls. It is a hovering abstraction, gliding above the ground. The building is not only in the landscape, but is the landscape, reflecting the greenery, shimmering like the water, flickering like clouds. It blurs and unifies, making the museum a park, the park a museum.

博物馆陈列展品，就像一个用来装满绘画作品的空箱子。这是一个美丽而纯净的大箱子，展览空间中除了一个脚手架以外再没有其他装饰。设计师要完成的是开放的、极具潜力的公共建筑，他们小心地建造引人瞩目的空间，避免将建筑视为孤立对象，而是将形式结合空间与光影来表现。

建筑底部的玻璃基础使建筑整体像一个漂浮于公园之上的珠宝盒。悬挑的雨棚和回墙形成围合之势。这是一个表面光滑的抽象几何体。建筑不但很好地融入周边风景，并且创造出更绚丽的风景。它的表面反射着绿草，闪耀着水光、映衬着白云，使博物馆与公园交融，你中有我，我中有你。

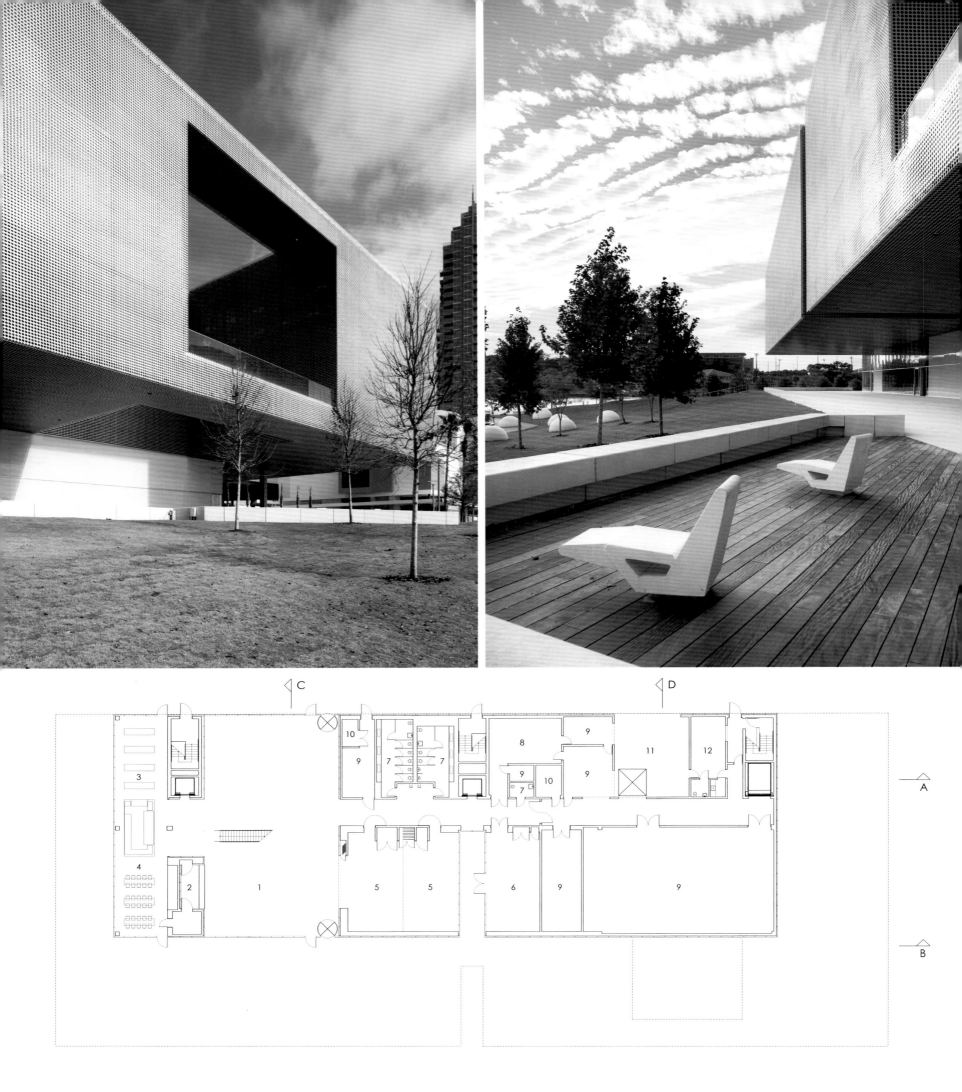

GROUND FLOOR   0' 10'   50'

1. LOBBY AND RECEPTION
2. TICKET DESK
3. MUSEUM STORE
4. MUSEUM CAFE
5. CONFERENCE ROOM
6. CLASSROOM
7. RESTROOM
8. CATERING KITCHEN
9. STORAGE
10. MAINTENANCE
11. LOADING DOCK
12. SECURITY
13. GALLERY
14. BALCONY
15. ATRIUM
16. WORKSHOP
17. STAGING
18. REGISTRAR
19. RECEPTION
20. OFFICE
21. KITCHEN
22. BOARDROOM

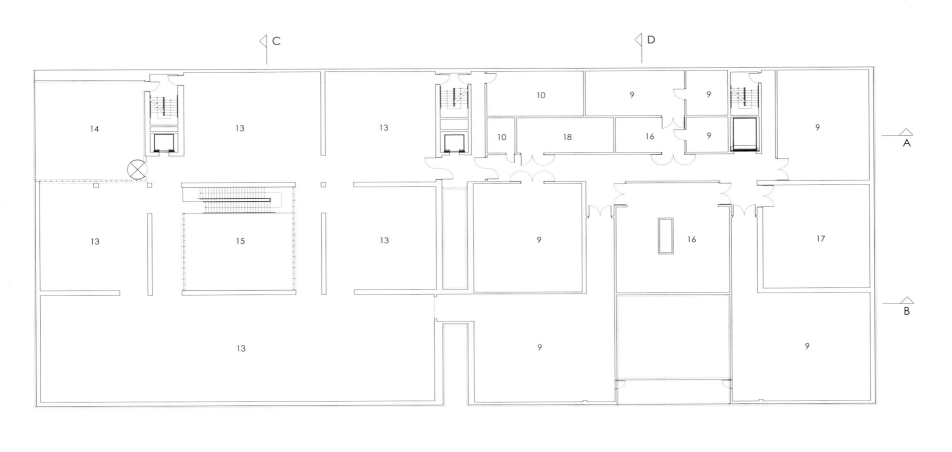

SECOND FLOOR

1. LOBBY AND RECEPTION
2. TICKET DESK
3. MUSEUM STORE
4. MUSEUM CAFE
5. CONFERENCE ROOM
6. CLASSROOM
7. RESTROOM
8. CATERING KITCHEN
9. STORAGE
10. MAINTENANCE
11. LOADING DOCK
12. SECURITY
13. GALLERY
14. BALCONY
15. ATRIUM
16. WORKSHOP
17. STAGING
18. REGISTRAR
19. RECEPTION
20. OFFICE
21. KITCHEN
22. BOARDROOM

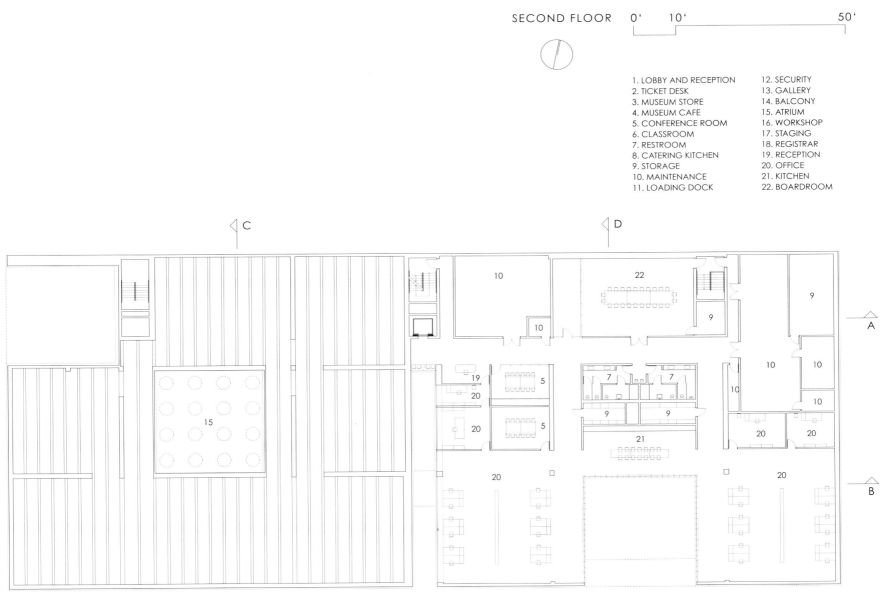

THIRD FLOOR

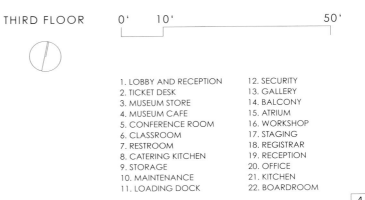

1. LOBBY AND RECEPTION
2. TICKET DESK
3. MUSEUM STORE
4. MUSEUM CAFE
5. CONFERENCE ROOM
6. CLASSROOM
7. RESTROOM
8. CATERING KITCHEN
9. STORAGE
10. MAINTENANCE
11. LOADING DOCK
12. SECURITY
13. GALLERY
14. BALCONY
15. ATRIUM
16. WORKSHOP
17. STAGING
18. REGISTRAR
19. RECEPTION
20. OFFICE
21. KITCHEN
22. BOARDROOM

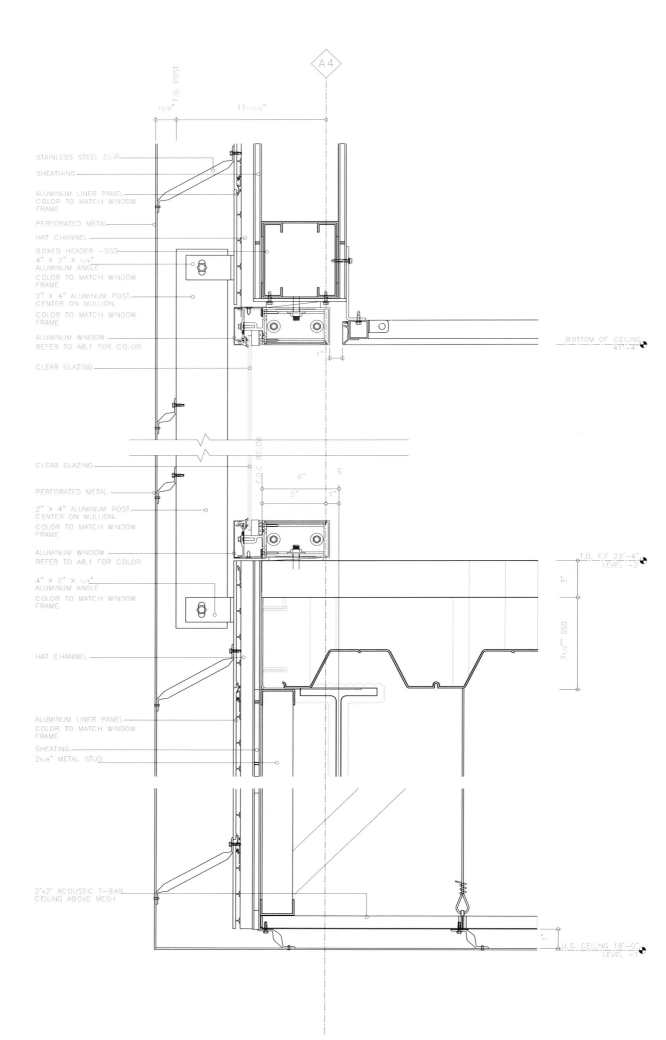

Edge Detail At Atrium Space

Shape and Layout   造型与布局

The long building is sliced in the center. This cut divides the programs in two, the one public and open, the other support and closed. Each of the two sections is organized around a court, one the lobby, the other a courtyard surrounded by the offices and curatorial areas. The 40' cantilever provides a huge public porch for the city, raising all the art programs above the flood plane. The walk along this porch, flanked by the park, focused on the river, leads to the lobby. The procession through this quiet and levitating space is the preparation for viewing art.

The lobby is at first horizontal, with entirely glass walls, two clear, two etched. The clear walls allow the site to run through the space, linking the Performing Art Building on the north with the turrets and domes of the University of Tampa on the south. Above the glass, the perforated ceiling wraps from the exterior into vertical perforated walls that turn into an upper ceiling, perforated again by a series of skylights. The galleries are reached from the lobby below via a dramatic cinematic stair reaching up. Below the stair is a bed of river rock. Off the lobby is a long glass room that houses the café and bookstore in a storefront along the riverwalk.

We have built the most expansive and generous field of galleries as instruments to enable, through curation, a world to expose art. They are arranged in a circuit, surrounding the vertical courtyard void. The galleries are blank, walls floor and ceiling all shades of white, silent like the unifying presence of snow. The floors are ground white concrete with a saw cut grid to echo the illuminated white fabric ceiling above. Linear gaps in the ceiling conceal sprinklers, air distribution and lighting. The second segment, around the open court, contains all the support for the museum. Offices surround the court on three sides. A bridge on the lower level is a secondary crossing from preparation to storage, a place for museum staff to be outside.

The image of the museum results from the nature of its surface - it does not symbolize or describe. It disengages through neutral form, providing a kind of pit stop in the attempt to represent. It is a moment to savor things in themselves.

By day the surfaces appear to vary almost, but never quite. They are smudged and stammering, with moray like images of clouds or water or vegetation, a shimmering mirage of reflections. It is an expansive and illusive image of a museum about things we don't quite know, about things we don't quite see. By day, light reflects on the surfaces. By night, light emanates from the surfaces. By night the exterior become a canvass for a show of light. The art form bleeds out onto the walls and escapes into the darkness. By night it is the magical illumination of the skin changing colors and patterns in endless variations which turn the building inside out, revealing it secrets as it broadcasts light, color and form into the city, duplicated in its reflection in the water.

This museum is both timeless and of our time, an electronic jewel box, floating on a glass pedestal, a billboard to the future, and a container to house works inspired with vision and able to show us other ways to see our world. The museum hovers in the park, a hyphen between ground and sky.

长条的建筑从中间被切开，分隔成为公共区与办公区，两个区域都围绕中心展厅布置。一个室内庭院环绕着办公区和馆长办公室。长12.6m（40英尺）的悬臂为城市提供了一个巨大的公共走廊，从水面上承托起所有的艺术珍品。走廊直通休息大厅，漫步其中如同漫步于水波之上、美景之中。参观者可以在这漂浮在空中的平和空间中尽情欣赏其中的艺术品。

休息大厅位于建筑首层，四壁由玻璃幕墙围合，两面是透明玻璃，另两面是磨砂玻璃。从透明的玻璃幕墙人们可以直接看到北边坦帕大学校区南边表演剧场的角楼和拱形穹顶。玻璃墙外带有空洞的天花板和墙面浑然一体，这些空洞同时也构成一系列采光孔。画廊从一层休息大厅经过一条奇幻的楼梯逐渐抬高，台阶下河水涓涓流过。休息大厅外是一个长长的毗邻河岸的玻璃廊道，两边布满了咖啡厅与书店。

设计者建造的是最宽敞、容量最高的艺术展览空间。展品环绕着建筑中的庭院合围一圈。博物馆内墙面、地面以及天花都是纯净而统一的雪白色。地板是一块块打磨过的白色混凝土，反射着雪白的屋顶。喷淋、通风口以及灯饰隐藏在屋顶板块与板块的缝隙中。围绕着室外花园的是后勤区，被办公室三面环绕，次要出口是一个位于地下室的桥梁，从准备间连接到仓库，作为博物馆工作人员的出入口。

博物馆的气质源于它自然的外形，没有那些浮夸的符号与装饰。中性的建筑形式具有典型休憩场所的特征，形成一个品味自我的圣地。白天建筑表面斑驳地反射着天空的云彩、荡漾的水波和公园里郁郁葱葱的绿植，如同空中的海市蜃楼般变化莫测。在这样一种若隐若现中，博物馆更显得引人入胜、扑朔迷离。白天反射光线，夜晚放射光线。夜晚，建筑开始讲述光影的故事。博物馆内的艺术品透过灯光映衬于夜空，通过不断变化的梦幻般的照明，将建筑的内外展现得淋漓尽致，宣扬着它的形式与色彩，在湖面形成粼粼波光。

博物馆既是永恒的也是现代的，一个电子的珠宝盒，漂浮在玻璃基础之上；一个未来的宣传册；一个装满人工雕琢器物的集装箱，激励我们从另一个视角观察世界。博物馆好似盘旋于公园内，连接天空与大地。

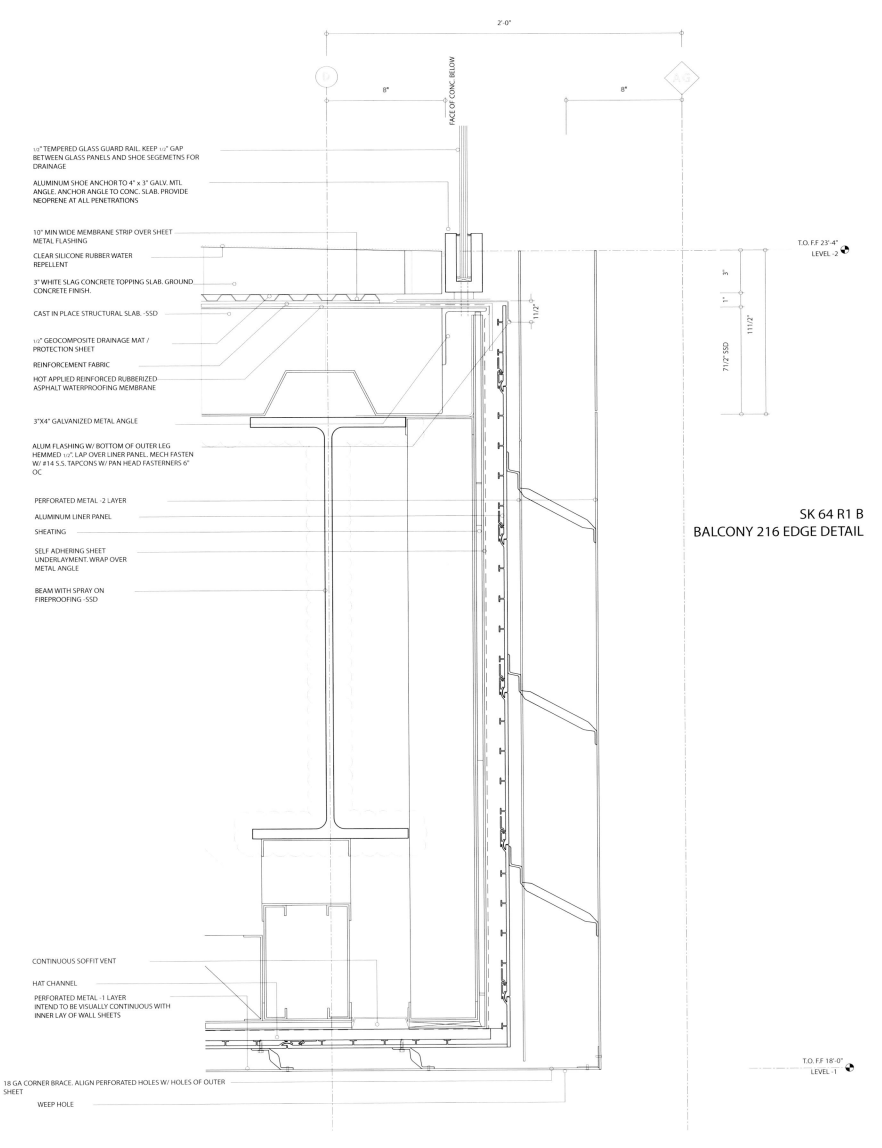

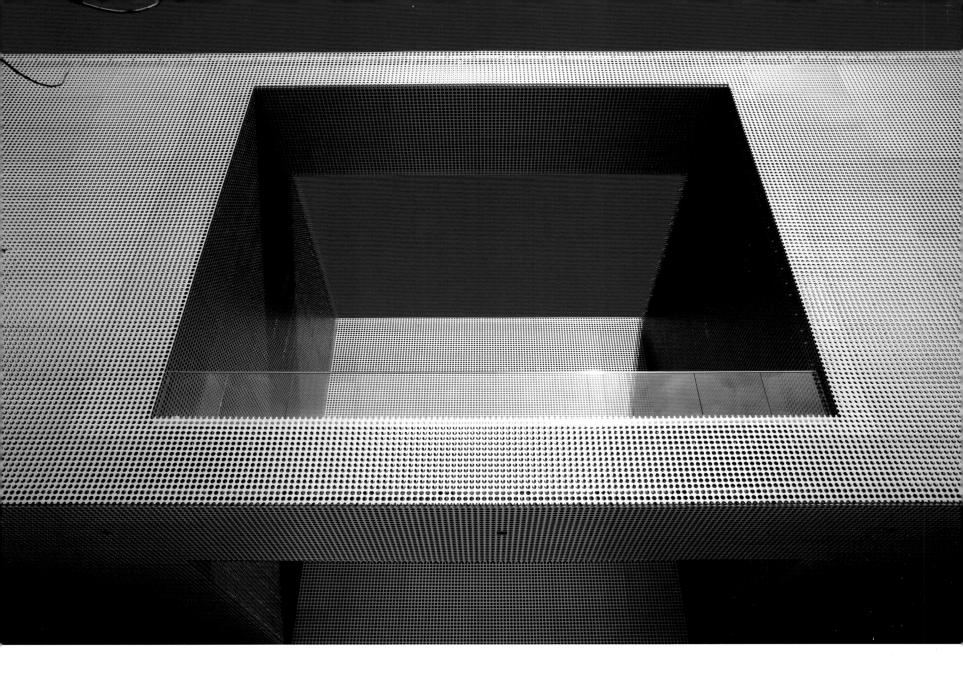

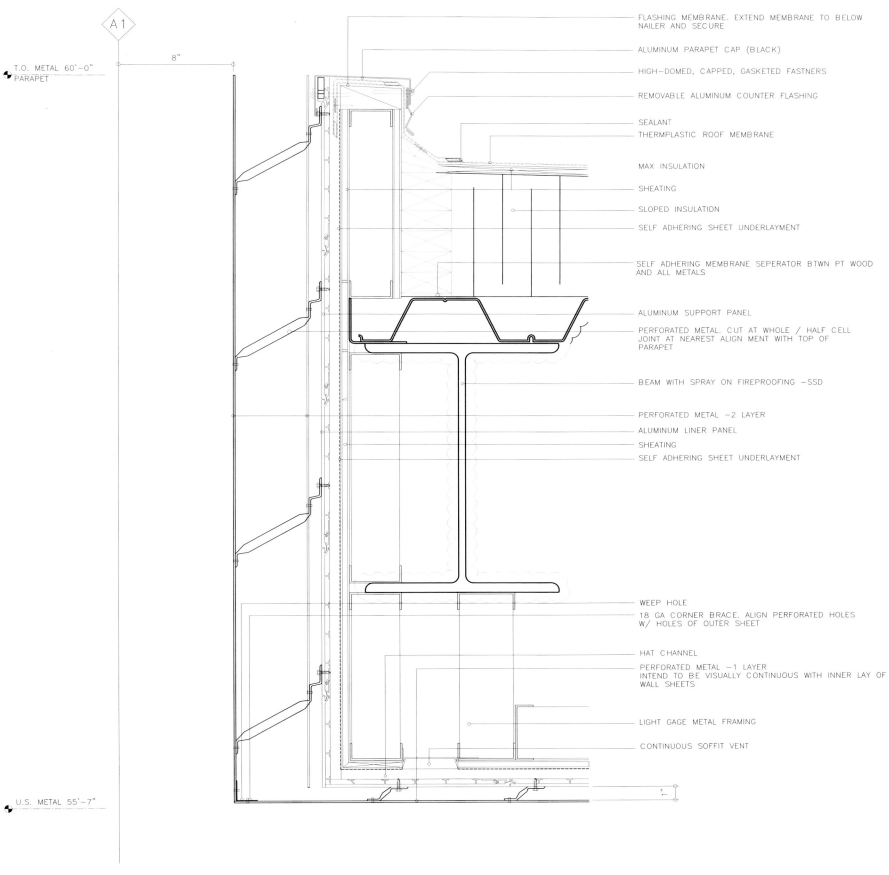

Parapet Detail At Balcony

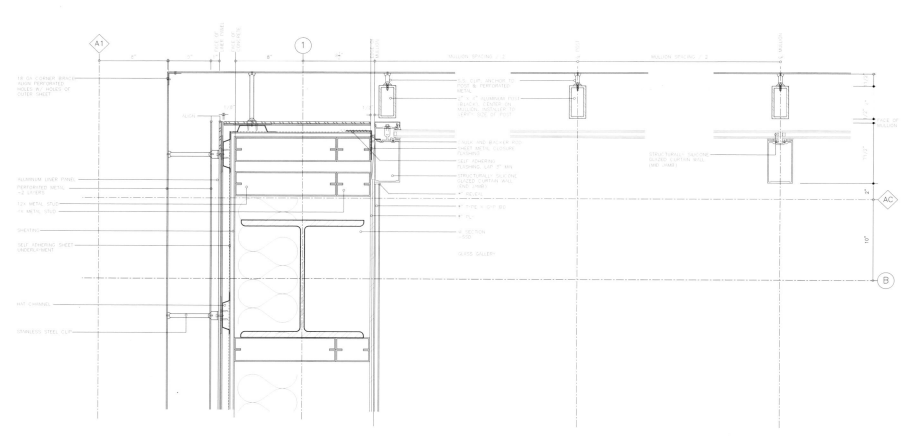

Curtain Wall Corner

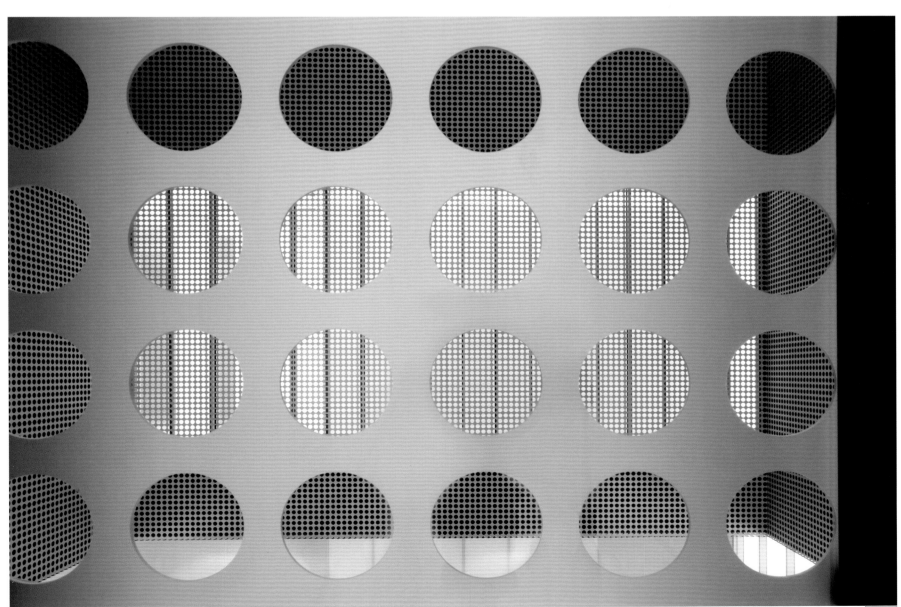

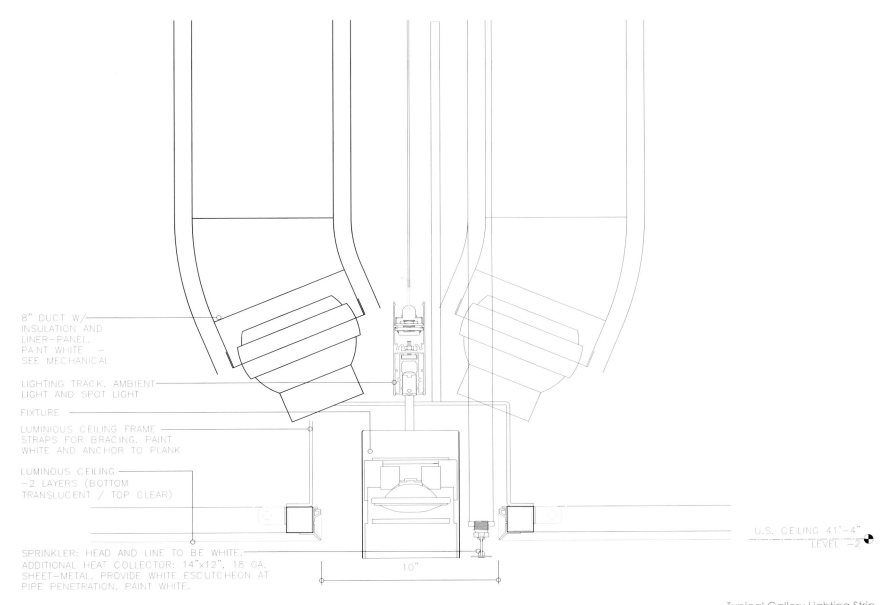

Typical Gallery Lighting Strip

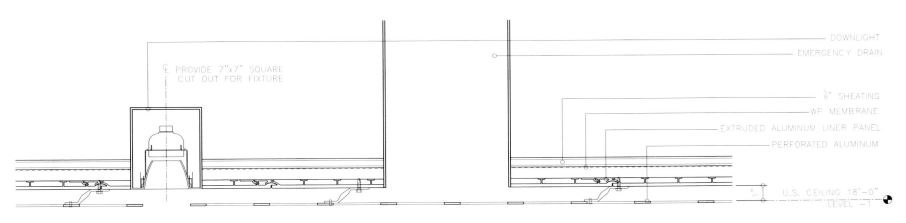

Typical Exterior Soffit

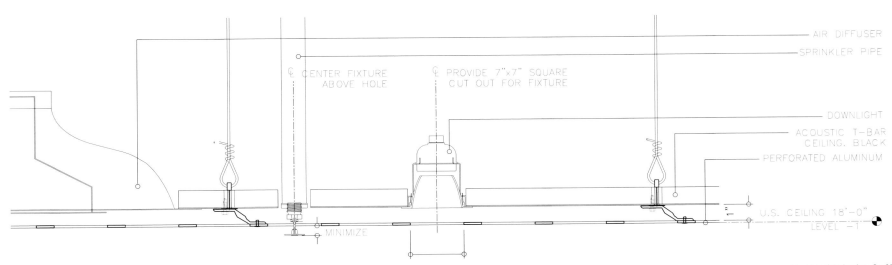

Typical Interior Soffit